More praise for Ethical Mark

"Hazel Henderson has been pointing the way to a responsible future for decades, and with this book she now shows us that, lo and behold, the future is here!"

—James Gustave Speth, Dean, School of Forestry & Environmental Studies, Yale Universoty

"There are thinkers, there are philosophers, there are prophets, and then there is Hazel Henderson. She sees the future and finds its seeds in the present. Visionary and practical, this book will make you feel better about the choices you make on behalf of humanity."

—Jessica Lipnack and Jeff Stamps, coauthors of *Virtual Teams* and *The Age of the Network* (www.netage.com)

"Hazel Henderson has compiled the stories of key players both within and without the establishment who are giving voice to an awakening environmental consciousness and the need to green the economy."

—David C. Korten, author of *The Great Turning* and *When Corporations Rule the World*

"Once again, it takes Hazel Henderson to reveal the truth the mass media ignores: this time that local and global economies are steadily shifting their goals and practices toward ethics, human well-being and sustainability. As socially and environmentally responsible investment grows, often pushed by civic action and with women playing ever more significant roles, we are steadily gaining ground in clean renewable energies, recycling, truly healthy food, and wellness production. If your local TV stations are not yet showing Henderson's inspiring *Ethical Markets* TV series, this book will make you demand they do!"

—Elisabet Sahtouris, PhD, evolution biologist, author of *A Walk Through Time*

"*Ethical Markets* is a fascinating guided tour through the emerging world of business as it can and should be. It reveals a parallel economic universe where authentic human needs are being met. *Ethical Markets* should be read by everyone who cares about our children and our planet."

—Riane Eisler, author of *The Chalice and The Blade* and *The Power of Partnership*

"Hazel Henderson has come forth with a magnificent map for action at the precise moment in our evolutionary crisis when we must redirect our most basic behavior. *Ethical Markets* is a must read."

—Barbara Marx Hubbard, author of *Conscious Evolution* and president, Foundation for Conscious Evolution

"Hazel Henderson's new book points to fundamental equations all of us—as leaders and strategists operating in every walk of life—must address as central to the creation of a healthy future for humankind. More than just an outstanding contribution to a deeper understanding of social and ecological responsibility, this book is a road map for strategists. This provocative book helps us explore innovative pathways toward a future that could be profoundly different from the world of contradictions we have built in the last several hundred years. Here, Hazel is at her best ever. She expresses a depth of wisdom that only comes from close contact with the deepest levels of global reality and the subtle games that are played therein."

—Oscar Motomura, CEO, Amana-Key Group

"This book is as thought provoking as its title. Henderson provides a comprehensive review of leading-edge global initiatives to sustain the planet, people, and profits. As a pioneer activist, Henderson has both authority and intimate knowledge to document current efforts to sustain humanity and the quality of life. The book makes an invaluable contribution to its subject by providing examples and references for others to promote, adopt and/or develop the green economy.

—Dr. Shann Turnbull, Principal, International Institute for Self-governance

Ethical Markets

Other books by Hazel Henderson:

Planetary Citizenship (coauthor 2004)

The Calvert-Henderson Quality of Life Indicators (coauthor 2000)

Beyond Globalization (1999)

Building a Win-Win World (1996)

Paradigms in Progress (1995)

The Politics of the Solar Age (1981, 1988)

Creating Alternative Futures (1978, 1996)

Ethical Markets

Growing the Green Economy

Hazel Henderson

With Simran Sethi

Foreword by Hunter Lovins

Chelsea Green Publishing Company
White River Junction, Vermont

Editor: Mary Bahr
Managing Editor: Marcy Brant
Project Editor: Collette Leonard
Proofreader: Lori Lewis
Indexer: Peggy Holloway
Designer: Robert A. Yerks, Sterling Hill Productions
Design Assistant: Peter Holm, Sterling Hill Productions
Printed in the United States
First printing, December 2006
10 9 8 7 6 5 4 3 2 1

Our Commitment to Green Publishing

Chelsea Green sees publishing as a tool for cultural change and ecological stewardship. We strive to align our book manufacturing practices with our editorial mission and to reduce the impact of our business enterprise on the environment. We print our books and catalogs on chlorine-free recycled paper, using soy-based inks, whenever possible. Chelsea Green is a member of the Green Press Initiative (www.greenpressinitiative.org), a nonprofit coalition of publishers, manufacturers, and authors working to protect the world's endangered forests and conserve natural resources. *Ethical Markets* was printed on Enviro 100, a 100 percent post-consumer-waste recycled, old growth-forest-free paper supplied by Maple-Vail.

Library of Congress Cataloging-in-Publication Data

Henderson, Hazel, 1933-
Ethical markets : growing the green economy / Hazel Henderson with Simran Sethi ; foreword by Hunter Lovins.
p. cm.
Includes bibliographical references.
ISBN-13: 978-1-933392-23-3
1. Sustainable development. 2. Informal sector (Economics) 3. Environmental responsibility. 4. Renewable energy sources—Economic aspects. 5. Quality of life. 6. Investments—Moral and ethical aspects. I. Sethi, Simran. II. Title. III. Title: Green economy.

HD75.6.H459 2006
338.9'27—dc22

2006026618

Chelsea Green Publishing Company
Post Office Box 428 · White River Junction, VT 05001
(802) 295-6300 · www.chelseagreen.com

To Brendan Alexander Cassidy, with love.

Contents

Foreword

ETHICAL MARKETS is a pioneering venture in what will soon be seen as the largest and most important movement in human history, the effort to devise a way of living on earth that simultaneously enhances human well-being and ecological integrity. Hazel Henderson has been one of the leading thinkers of this movement since its most recent inception in the early 1970s. One could well argue, however, that this has been a strain of human thought going back to the earliest religious leaders, the natural philosophers, and certainly such writers as Thoreau, John Muir, Aldo Leopold, and more recently Rachel Carson, David Brower, and Dana Meadows.

Hazel differs from these luminaries chiefly in that, happily, she is very much alive, and she is still creating innovative ways to carry the word to ever-broader audiences. But she is every bit their equal in stature and importance of her contribution.

The movement is perhaps even more important these days, when it has become clear that humans have the capacity to destroy whole environments. This is a relatively new realization. In the 1930s, the physicist Robert Milikin opined, "There is no risk that humanity could do real harm to anything so gigantic as the Earth." Ironically, that was the same year that the engineer Thomas Midgely invented chlorofluorocarbons, the chemical that has been chewing a hole in the planet's stratospheric ozone layer.

In 2005 the Millennium Ecosystem Assessment, a report by 1,360 scientists from ninety-five countries around the world, stated that "human activity has polluted or over-exploited two-thirds of the ecological systems on which life depends." Human activity, the report stated, is putting such strain on the natural functions of the Earth that the planet's ability to support future generations can no longer be taken for granted. Kofi Annan reacted to the assessment's finding by saying "the very basis for life on earth is declining at an alarming rate."

Sobering words. What has brought us to this state of affairs? And what are we to do about it? The primary instrument of the destruction is business as it seeks to meet our increasingly rapacious appetite for goods and services. A growing population becoming ever more affluent, at least in the West and in enclaves in every country, fuels the actions of companies. But business is also the most likely instrument of our delivery. Business leader Ray Anderson asked, "What is the business case for ending life on earth?"

Hazel is among the leaders of those proving that there is none, and that there is a very good business case for behaving in ways that not only generate profit but that also protect people and the planet. This formulation, in John Elkington's words, of the triple bottom line is increasingly giving way to the integrated bottom line, to the realization that in a world beset with challenges of the likes of peak oil, climate change, loss of species, destruction of ecosystems, and rampant social inequities, companies that behave responsibly to all living things will enjoy increased profits because of this behavior. As Hazel shows in these stories, the reasons are varied. To start with, companies who manage to do no harm, and in fact serve life, manage to catch the sorts of irresponsibility that resulted in massive shareholder dissolution at Enron. They don't make the sorts of costly mistakes that result in pollution. They don't anger stakeholders, who, in an Internet-empowered world, can deny a company the franchise to operate.

Second, companies that manage responsibly reduce their use of resources because it is the right thing to do. But they also enjoy cost savings and increased brand equity. Such behavior motivates their workforce and they achieve higher productivity, and perhaps more importantly, companies in turn attract and retain the best talent.

In a carbon-constrained world, reduction of waste cuts costs and vulnerabilities. Swiss Re, the major European reinsurer, recently stated to its customers that if they do not take their carbon footprint seriously, perhaps Swiss Re will not wish to insure them—or their officers and directors.

For years a small nonprofit in the United Kingdom, the Carbon Disclosure Project, has sent out a questionnaire to the Financial Times 500 (the five hundred biggest companies in the world) asking companies to disclose their carbon footprint. For almost a decade, fewer than 10 percent of the companies answered. In 2005, 60 percent replied. This was in part because the project now represents institutional investment funds with over $31 trillion in assets. In addition, under the Sarbanes-Oxley legislation, managers who fail to disclose information that could materially affect share value can be personally criminally liable. So, what's your carbon footprint? In early 2006 the fund sent its questionnaire to the largest eighteen hundred companies in the world. Given the project's financial backers, these are inquiries that will be hard to refuse.

Companies that undertake an aggressive carbon reduction plan are finding dramatic savings (DuPont, which has already met its 2010 target of 65 percent reductions of emissions, reckons to save $2 billion), and that such programs drive their innovation and increase market share (ST Microelectronics found that such efforts took their company from the number twelve chipmaker in the world to the number six, with projected savings of almost a billion dollars). Some financial advisors already state that a commitment to sustainability is the hallmark of good corporate governance and the best indicator of management capacity to protect shareholder value. Indeed, the Dow Jones Sustainability Index outperforms the general market, and the Domini Index of Socially Responsible Companies has outperformed the Standard & Poor's companies for over a decade.

In the spring of 2005, the socially responsible investment research firm Innovest Strategic Value Advisors released a report showing that in whole industry sectors, from forest products and paper to oil and gas and electric utilities, the environmental leaders in the sector are outperforming the environmental laggards.

A corporate commitment to sustainability enhances every aspect of shareholder value. As the examples above show, using resources more efficiently, redesigning products in ways that mimic nature, and managing to restore and enhance human and natural capital offer ways to:

Reduce costs and increase profitability and financial performance

Reduce risk

Retain the franchise to operate and reduce legal liabilities

Attraction and retention of the best talent

Enhance ability to drive innovation

Labor productivity—increased worker health

Increase market share—enhanced brand equity

Product differentiation

Ensure supply chains and stakeholder management

Taken together, the elements of sustainability confer the ability to be "first to the future." The companies that practice this approach will be the billionaires of tomorrow.

In May 2005 Jeffrey Immelt, the man who replaced Jack Welch at the helm of General Electric, stood with Jonathan Lash, the president of World Resources Institute, a leading environmental organization, to announce the creation of GE Ecomagination. Immelt committed the only company, which had there been a Fortune 500 around in 1900 would still be on it, to implement aggressive plans to reduce emission of greenhouse gases. In a joint article in the *Washington Post,* and subsequent ads purchased in major sporting events, GE trumpeted its commitment to behave more sustainably.

The point is not so much the substance of the announcement, which amounts to less than companies like DuPont (whose senior management was understandably annoyed that the newcomer was getting all this ink) actually do on a daily basis, and have, under Chad Holliday's leadership, for over a decade. Rather the announcement signaled a tipping point. How do you carve out a legacy when you follow the man widely credited with saving the company? You announce you are going to save the world. This book profiles the movement, now millions strong around the world, of companies and citizens who are doing just that. You can, too.

L. Hunter Lovins
President, Natural Capitalism Solutions
Chair, Sustainability Strand
Presidio School of Management

Introduction

THIS BOOK IS ABOUT the cleaner, greener, more ethical, and more female sectors of our U.S. economy—and many others around the world. These growth sectors can employ every man and woman able to work and are the key to a sustainable and healthy future for humanity. These segments of the business market are here today and have been quietly growing for over twenty-five years, yet virtually ignored by mainstream financial media. How could this have happened? Why did it take until 2006 for a U.S. president to finally admit that the country is addicted to oil? I explored these issues in *Politics of the Solar Age* (1981, 1988) in the hope that the transition from fossil-fueled industrialism to renewable energy and sustainable technologies would begin in the 1980s. I outlined the case for this inevitable transition and pointed to all the emerging science and technologies to achieve this design revolution. I warned that all the entrenched industries based on fossil fuels, the powerful interest groups, and the conventional consumerism of commercial mass media would stand in the way. I even cited all the evidence of impending environmental depletion and pollution, and the faulty economics that blinded policymakers and private decisions. I failed to realize, however, in my optimism that full systemic social change would take another generation. Nonetheless, in spite of the blindness and incomprehension of mass media, the new "sustainability" sectors began to emerge in many countries.

Changes toward a green economy can be grouped into three main areas:

1. **The LOHAS (lifestyles of health and sustainability) sector**: renewable energy and resource industries (solar, wind, biomass, oceans, hydrogen, fuel cells, etc.), those in recycling, remanufacturing, reuse, barter, and secondhand auctions (like eBay), those in preventive, alternative healthcare, wellness, fitness, etc., and those companies in clean food and organic agriculture (www.lohas.com)
2. **Socially responsible investing**: the fastest growing segment of U.S. capital markets (representing about one in every $11 invested in publicly listed companies) or some $2.3 trillion, according to the Social Investment Forum (www.socialinvest.org)
3. **Corporate social responsibility**: Management is increasingly focusing on their own social responsibility. Most global companies have forsaken orthodox economic ideologies of laissez-faire (unregulated markets), famously promoted by University of Chicago economists including Milton Friedman, that "the only business of business is to make profits for shareholders." This view that relies on markets as self-correcting holds that government regulations are ineffective, self-defeating, and usually unnecessary. History has already overtaken these views. Companies today acknowledge that globalization of information technologies has morphed into a new age of truth. No corporate activity, which may affect society and other stakeholders (employees, suppliers, customers, host countries, and the environment), remains unnoticed. Thousands of civic groups, like Corpwatch.org, Global Exchange, the World Social Forum, and many focused on specific issues from GMO-foods to global warming, monitor every corporate move. Their Internet reports and blogs can break a precious corporate brand and stock prices in real time. Thus, corporate CEOs today have installed a myriad of in-house programs, often personally overseen by vice presidents of corporate responsibility. These new activities include hewing to new standards from ISO 14001 and EMAS

to SA-8000 and many other "good citizenship" accreditations, labels including the U.S.A.'s Green Seal, Germany's Blue Angel, and many others. In a 2005 poll of CEOs by the World Economic Forum and KPMG, 70 percent said that "good corporate citizenship" "was vital to profitability." In 2006 the EthicMark I created became available to recognize advertising and media campaigns that uplift he human spirit and society.

So it becomes clearer why these three burgeoning sectors of the U.S. and global economy have been all but invisible on mainstream financial media: a fierce paradigm war of worldviews is underway. Most media outlets are owned by only a few, yet very large, conglomerates: News Corp, Time-Warner, Disney, GE, VIACOM, and others. These companies are deeply embedded in unsustainable, wasteful, fossil-fueled, and nuclear-powered sectors of the global economy, along with global banks and firms that finance their expansion. I described this new form of government, "mediocracy," in *Building a Win-Win World* (1996). These three emerging sectors, which we cover in the *Ethical Markets* TV series and in this book, pose a direct challenge to the market dominance of the existing world economic players. No wonder reporting on the explosive growth of these new sectors is sparse. Those who discuss them are often trivialized as unrealistic, diehard hippies from the 1960s, and the start-up companies are seen as anomalies. Even many socially responsible funds with stellar performance (often outperforming mainstream portfolios) have been denied or disparaged. For decades, many brokers would tell their clients that they would lose their shirts investing in these funds. Even interviewers who are as skilled as Charlie Rose, Bill Moyers, and Lou Dobbs are still too immersed in the dominant paradigm to reject these fallacies. Getting hidden assumptions out onto the table not only enriches the discussion but can relieve much social distress and lead toward more realistic debates on public policy and priorities.

Our TV series was created to cover these three cleaner, greener, more transparent, and ethical parts of the economy, many of which are spearheaded and led by women. Along the way, it was necessary to unravel much of the now clearly obsolete thinking based on eighteenth and nineteenth century economic ideas—deep in textbooks and computer models. The word is out that economics, never a science, has always been politics in disguise. I have explored how the economics profession grew to dominate public policy and trump so many other academic disciplines and values in our daily lives. Economics and economists view reality through a monetary lens. Everything has its price, they believe, from rain forests to human labor to the air we breathe. Economic textbooks, Gross National Product (GNP), and the statistics on employment, productivity, investment, and globalization—all follow the money. Happily, all this focus on money exposes how money is designed, created, and manipulated. Our widespread focus on the politics of money is at last unraveling centuries of mystification.

Civic action with local currencies, barter, community credit, and the more dubious rash of digital cybermoney reveal the politics of money. Traditional economics is now widely seen as the faulty source code deep in societies' hard drives, replicating unsustainability: booms, busts, bubbles, recessions, poverty, trade wars, pollution, disruption of communities, and loss of culture and biodiversity. Citizens all over the world are rejecting this malfunctioning economic source code and its operating systems such as the World Bank, the International Monetary Fund (IMF), the World Trade Organization (WTO), and imperious central banks. Its hardwired program—the now derided "Washington Consensus" recipe for hyping GNP growth—is challenged by the Human Development Index (HDI), Ecological Footprint Analysis, the Living Planet Index, the Calvert-Henderson Quality of Life Indicators, the Genuine Progress Index, and Bhutan's Gross National Happiness, not to mention scores of local city indices such as Jacksonville, Florida's Quality Indicators for Progress, pioneered by the late Marian Chambers in 1983.

As with politics all real money is local, created by people to facilitate exchange and transactions, which are based on trust. Events of the past twenty years have necessarily recast the story of how this useful invention, money, grew into abstract national fiat currencies backed only by the promises of rulers and central bankers. We witness how information technology and deregulation of banking and finance in the 1980s helped create today's monstrous global casino where $1.5 trillion worth of fiat currencies slosh around the planet daily via mouse clicks on electronic exchanges—90 percent in purely speculative trading.

In view of these abuses, the task before us is nothing less than to redefine success, wealth, and progress for our massively changed circumstances in this twenty-first century. Our human family now has over six billion members and we consume 40 percent of all the primary photosynthesis-based production of the planet's plants on which we depend. The biosphere and its thin layer of biodiverse life forms are now suffering their next biggest extinction in history. Even once-skeptical scientists confirm that burning fossil fuel, and other human activities, causes global warming, which has played a role in recent weather disasters: floods, droughts, and stronger, more frequent hurricanes. *TIME*'s cover story (Mar. 27, 2006) and former Vice President Al Gore's movie *An Inconvenient Truth* at last broke the mass media's avoidance of global warming—calling for immediate changes to limit CO_2 emissions. Even Wall Street is learning that "business as usual" is no longer an option. Hedge funds and pension plans are investing in ever-riskier asset classes such as catastrophe bonds—betting against insurance losses mounting rapidly from such natural disasters. Colin Challen, chair of the All-Party Parliamentary Climate Change Group in Britain, in a speech on March 28, 2006, called for the Contraction and Convergence plan of the Global Commons Institute, based in the U.K. (www.gci.org.uk), which calls for globally shared "emission rights" for every man, woman, and child, so that poorer people could sell theirs to the richer—thereby converging on equitable reductions of CO_2. This emissions-trading approach is similar to the proposal by Argentinean economist and mathematician Graciela Chichilnisky, the inventor of "catastrophe bonds," for an International Bank for

Environmental Settlements to administer such a global approach to equitable reduction of CO_2. Giving rights to pollute to corporations was inequitable and most are now auctioned.

Thus, chapter 1, "Redefining Success," is an introduction to all these new views, sensibilities, lifestyle changes, employment and career goals, and investment strategies. We look at the new ways of measuring success, wealth, progress, productivity, efficiency, and so forth, and the broader, multidisciplinary indicators emerging at all levels of societies worldwide that are helping steer humanity to a brighter future. This new worldview underlies the strategies of the companies we feature in this book, who employ the new "triple bottom line" accounting (people, planet, profit), and the eighty visionary CEOs and leaders we interviewed.

After redefining the terms for success we discuss "Global Corporate Citizenship." The globalization of finance and technology and the increasing influence of global corporations now challenges the sovereignty of even the most powerful nations. The uproar in the United States over who should own, operate, and control U.S. ports and other basic infrastructure was emblematic of the growing debate over globalization. Outsourcing has become another flashpoint, as is immigration. Mainstream media pick up on these symptoms—but often ignore the deeper implications of current forms of globalization of technologies. Free market ideas about the merits of deregulation, privatization, and world trade are promoted as a win-win for all. These obsolete economic textbook formulas have caused widespread social, cultural, and environmental disruption. Today, these largely U.S.-driven globalization effects are finally being felt in the U.S. homeland. Yet we saw few commercial mass media stories attempting to unravel and analyze the globalization phenomena—due to shrinking foreign bureaus, shorter news, and sound byte coverage in the drive for profitability. Yet, all these trends affect the lives of people and countries—for better or worse. As we see from the blogosphere, there is worldwide, grassroots concern over corporate responsibility regarding human rights, workplace safety, decent wages and conditions, and protection of the environment. While the corporate scandals continue to mount, employees lose their jobs, life-savings, and pension plans are in jeopardy. We interview leading socially responsible investors, managers of mutual funds, pensions, and church and university endowments who offer safer, healthier alternatives. They have joined civic groups, labor unions, women, environmentalists, and those concerned with social justice and human rights in calling for greater corporate accountability—particularly after their heavy losses in Enron and other companies. We describe "involuntary investors" (a significant percentage of the hundred million U.S. adults, who are invested in the stock markets) and present better ideas for their pension plans, 401Ks—which have recently lost so much value. These involuntary investors are calling for more transparent, ethical corporations that can restore their trust and promote their goals and values. We show how all the new initiatives in corporate social responsibility, socially responsible investing, and the LOHAS sector, as well as the more than three thousand companies who have signed on to the United Nations Global Compact (described in chapter 2) are responding. We began to see the evolution of capitalism itself. More ethical markets are now necessary in the twenty-first century information age—now morphing into that new age of truth as global public opinion becomes the world's newest superpower. Markets can only operate where there is trust, transparency, honesty, and fidelity in contracts—as well as service to customers and other stakeholders in society. The leaders we met understand the risks of corporate misbehavior to their revered brands and reputations. As a result, they are embracing social, environmental, and ethical auditing, the new scorecards and indexes, which redefine success, progress, efficiency, and productivity in the wider social context.

Most people are unaware that all the financial and business news, economic policies at state, local, and national levels—in all countries—are based on economic statistics that reflect only one half of the full range of production, services, investment, and exchanges in societies;

only half of which are conducted in money. The equally important nonmoney sectors (in many countries much larger than the official, money-denominated sector traced by gross national product (GNP)/gross domestic product (GDP) and other macroeconomic measures) are in reality the core platform of social life. These non-monetary contributions form what we call the Love Economy—the families, communities, cooperatives, and voluntary sharing activities that underpin the competitive money-based sectors. We introduce some of its most inspiring leaders. Unless policymakers and the public are clear about the vital role of this Love Economy, it is devalued, shortchanged, and begins to crumble. Volunteers begin to disappear as they join the paid labor force. Economists erroneously categorize homemakers, stay-at-home moms and dads, as "not economically active."

The textbook model of human nature in economics is the "rational economic man" who maximizes his own self-interest in competition with all others. In reality we know that people are also equally cooperative and enjoy sharing and giving. New brain science and microbiology now show that economics is founded on a set of core assumptions that are invalid (see www.hazelhenderson.com). For example, the Washington-based nonprofit group Voluntary Sector has estimated that over eighty-nine million Americans volunteer at least five hours per week to their communities. In 1995 the United Nations Human Development Report found that unpaid work and production of goods and services was equivalent to $16 trillion—simply missing from the official global GDP figure of $24 trillion. So two-thirds of the world's product went uncounted, unrecognized, and unappreciated. Such enormous value to societies represents a vast appreciating asset! Futurists Alvin and Heidi Toffler describe its dimensions in *Revolutionary Wealth* (2006). Similarly, nature's wealth goes uncounted in GNP. New broader statistics on health, education, social capital, and ecological assets are now creating better scorecards of wealth and progress—beyond money and GNP. Life is rich in many dimensions and we know that money can't buy many of the things we most desire—such as love and happiness.

Health is another universal aspiration. Green building and design offers hope of a healthier future built environment. An "efficient" building serves the broader bottom line—beyond money as covered in the PBS series *Design e-2*. Nature is often the unappreciated asset. Visionary architects, including Bill McDonough, are bringing free sunlight and fresh air back into offices. Increasingly, architects and planners are using rooftops to maximize energy efficiency and provide space for food production. As oil becomes more expensive and fossil fuels continue to pollute our atmosphere with CO_2, all the old metrics are changing.

Our quality of life has much to do with the vitality of the community in which we live. Healthy communities typically have stable families, enjoyable neighborhoods, and businesses that revitalize the local economy. Because economists have not measured the deeper, broader kind of efficiencies provided by cohesive communities and the values of families and local cultures, these local living economies have been undervalued—until they break down. Then social services, unemployment, drug and crisis counseling, and caring for homeless people require huge taxpayer costs. Today, many of the smartest investors, asset managers, and pension funds are joining with local leaders in reinvesting in these vital community redevelopment efforts as described by Michael Shuman in *The Small-Mart Revolution* (2006). A new breed of accountants and statisticians are finding the hidden wealth and opportunities for new jobs, businesses, and residential developments in local communities as they embrace new definitions of success, wealth, and progress.

A more dimensional understanding of these terms affects the global stage as well. Countries strive to be successful exporters. Economic textbooks claim that more trade—now worldwide—is good for everyone. The World Bank would advise countries how to "grow their economies" by exporting similar products to world markets, often glutting them. The World Trade Organization, set up in 1996, created its rules based on these assumptions. But the old textbook model of "free trade" assumed that capital stayed within countries' borders and that all the trading countries would benefit, even those who had little power and were way behind in industrial development. So the economists' recipes for economic development—measured as GDP growth—urged them to open their borders, reduce tariffs, make their currencies convertible, privatize their main industries, and allow foreign capital to flow in and out freely—this action is known as the "Washington Consensus." These measures work for a big successfully industrialized country, but it is now well recognized that less powerful developing countries can lose out, with their weaker, smaller companies, and farmers being put out of business.

Today, battles rage over all these issues, as currency trading and herds of electronic bulls and bears create gigantic waves of "hot" money that sloshes around the planet daily. Even the most democratic and competent leaders lose control of domestic affairs, while local people stage demonstrations to prevent their water supplies, national resources, and biodiversity from being bought-up and privatized. Since its failed Doha Round in July 2006, the WTO has become almost irrelevant. The new "resource-nationalism" is evident as China and India scour the world, buying up assets in the energy and resources sectors. In Latin America, resource nationalism has led to widespread rejection of the Washington Consensus formulas and a new group of leaders in Venezuela, Argentina, Brazil, Bolivia, Peru, Uruguay, and Chile are articulating more sustainable forms of development focusing on social equity and justice for their poorer, rural, and indigenous peoples. Argentina, Venezuela, and Brazil have now largely paid off their loans to the IMF to free themselves from Washington Consensus prescriptions and have saved several billions of dollars in interest payments.

Proponents of Fair Trade in many commodities support this new approach to development. Growing numbers of philanthropists, socially concerned investors, and entrepreneurs are creating healthier models of fair trade. They bring together local farmers and small producers to create eco-friendly, healthy products that benefit the local community. The new scorecards measuring wealth, progress, and quality of life are slowly steering policymakers at all levels to reevaluate which kinds of exports spread benefits more fairly. Subsidized tax-supported ports, transport, and facilities, as well as energy prices that ignore environmental and social costs underpin most conventional world trade today. If world trade were to fully account for these huge subsidies we would find that local and regional trade is more efficient. (More on these trends at www.Calvert-Henderson.com.)

Women own a great many fair trade and green businesses. In fact, women-owned businesses now represent some 50 percent of all privately held companies of all types—from construction and science to health care and environment. As they increasingly operate the world of commerce and industry, women are redefining success because their life goals differ from those of their male counterparts. Women business owners do not typically put making money at the top of their goals. Rather, they cite the need for personal autonomy and flexibility to manage their complex lives, the "glass ceiling" in so many corporations that limited their advancement; the satisfaction of personal creativity; and the freedom to address unmet needs with their business models. Women-owned and managed businesses now employ over nineteen million people. Women, as an enormous pool of natural assets, have been undervalued for decades. At last, society is beginning to appreciate their role in creating wealth and progress.

Many businesses, women-owned and otherwise, are leading the development of renewable energy. With advice from Amory and Hunter Lovins, John Todd, and other world-class experts, we examine the great transition from early industrialization based on fossil fuels to renewable forms of energy. The public debate about declining global oil production and global warming has at last begun. Rising oil prices have spurred the rethinking of U.S. energy futures and how to reduce our vulnerability to foreign supplies. In spite of analytical habits that overlook the full value of investments in all forms of renewable energy and the huge savings from the technologies of efficiency, entrepreneurs, technologists, the inventors, and venture capitalists are now breaking the logjams. A cleaner, greener, more energy efficient future is coming into view. Europe, China, and Japan are streaking ahead of the United States in solar and wind power. China and India have much to teach us about traditional ways of meeting human needs while preserving natural resources and the environment. The Midwest has been described as our "OPEC of Wind Energy" and offers the possibility of meeting much of our own domestic electricity needs. The dimensions of this great transition to the solar age cover every sector of our economy: agriculture, construction, urban design, transportation, infrastructure, chemicals, and pharmaceuticals, as well as the new smart electricity grids. This energy transition can create millions of new jobs, clean our air and water, and reduce problems with CO_2 buildup and global warming. The Carbon Disclosure Project, representing 211 investment mangers with $31 trillion in assets, asks major corporations to disclose their CO_2 emissions and policies (cdproject.net).

One of the most surprising aspects of the new twenty-first century capitalism is the rise of concerned, active shareholders. They invest not only for economic returns but to help create a better world. You will meet some who attend annual company meetings and challenge management policies on a host of issues that concern them such as: fair treatment of employees, pollution, outsourcing jobs to low-wage countries, minority rights, diversity of boards and management, climate change, and corporate governance. The active shareholders we interviewed are also influencing investment choices of pension funds, university endowments, foundations, and socially responsible mutual funds. Controversial in the 1970s, shareholder activism is now popular and widely recognized as a progressive movement in the evolution to more ethical twenty-first century capital markets. Shareholders love the extra psychological "bang for their bucks," while the sheer power of the $2.3 trillion active U.S. investors wield is leading to a new model of the corporation managed not only to the benefit of shareholders but all stakeholders including employees, customers, suppliers, community, and the environment. Stakeholder capitalism is the wave of the future—thanks to the millions of shareholder activists in many countries helping change the game and the scorecards of social progress and human development.

Changes at the top are affecting workers as well. In traditional societies, work is still often unpaid in local villages and rural agriculture, where people grow their own food and build their own houses and community facilities in mutual cooperation. As the industrial revolution spread from Britain three hundred years ago, former peasants who used the village commons to graze their sheep and weave their wool and clothing in their cottages were replaced by enclosure laws and factories. Millions who were denied the use of formerly common lands became impoverished and hungry. Industrialism also unleashed incredible creativity and technological innovation. Industrialism entailed labor saving—doing more with machines and energy rather than human beings. As globalization accelerates, these technological changes continue to change our workplace, careers, and opportunities, and create new training and education needs. Outsourcing is accelerating in the United States, shrinking our manufacturing and service sectors. No longer can people expect to make a lifelong career with one company or one industry. Most of us expect to be lifelong learners, while many can now opt to be self-employed or entrepreneurs, due to information technologies. In this book we explore the good and bad news for individuals, businesses, communities, and countries as they navigate these global transformations and hear from thinkers like Jeremy Rifkin, Patricia Kelso, and CEOs of employee-owned companies.

The industrial revolution changed not just our work lives but the goods we buy, including food. Many scary stories over the past twenty-five years concerning the industrialization of our food supplies have led to the explosive growth of the clean food and organic agriculture industries. Fears of mad cow disease, the effects of eating genetically modified foods, and ingesting toxic pesticide residues, and the rise in childhood obesity have moved people to rediscover the health benefits of fresh, locally grown produce, free-range eggs, and foods free of allergy-causing additives and disruptive hormones. We solicit leaders from the pioneer

Rodale Institute and many whole food companies to comment on this burgeoning new market, growing at an estimated 20 percent per year.

The current trend in favor of locally grown and organic foods coincides with changes in attitudes toward health care. Industrialized medicine has reached a crisis with widespread dissatisfaction among patients, doctors, nurses, hospitals, and all aspects of today's medical-industrial complex. The United States spends more than any other country on medical costs per person—almost 16 percent of the GDP—with no better outcomes than countries spending one half of this amount. Yet, as *Business Week* pointed out in "What's Really Propping Up the Economy" (Sept. 25, 2006) this wasteful medical-industrial complex accounted for 1.7 million new jobs since 2001, while the rest of the private sector added few. The failure of our medical programs has given rise to a rapidly growing new sector of our economy—based on the philosophies of prevention and natural (and cheaper) approaches to wellness. Nowhere is redefining success more emblematic than in this sector, where less is more, and love and personal caring are valued over high-tech interventions.

Lastly, we cover the emerging trends in socially responsible investing. At last, mainstream venture capitalists are following the lead of the many pioneers who have been funding companies in solar, wind, biomass, fuel cells, hydrogen, and more efficient technologies of all kinds. Leaders from D. Wayne Silby, founder of the Calvert Group, Robert Shaw of Arete, and Nick Parker of Cleantech Ventures, convey their enthusiasm for continually seeding these sustainability companies, many of which are destined to become the "IBMs" and "Microsofts" of the twenty-first century. Even the once skeptical *The Economist* in its "Survey of Climate Change" (Sept. 9, 2006) covers the greening of business worldwide and the boom in clean technologies. In the same month, *Business Week* launched a new section called "Green Biz." In its October 16, 2006 issue, *Fortune* added a sixteen-page supplement "It's Good to be Green."

ONE

Redefining Success

WHAT IS SUCCESS? A Potlach Indian from the Northwest measures success by how much he or she can give away to help others. An average Wall Streeter might describe it in terms of rising stock portfolios, a bigger car, and swankier apartment. To an Amish farmer in Pennsylvania, success is a well built home, fertile fields, and independence from high technology and the evils of big city life. For a Chinese peasant, it's a low-paid job in a city factory. For most of us, a happy home and personal life are the basic hallmarks of success. As notions of success vary in our six-billion-plus member human family in today's global economy, so do the ways countries and corporations measure and define success, wealth, and progress.

Every day, hundreds of thousands of reports in all media discuss the status of the Gross National Product (GNP) and its narrower domestic version the Gross Domestic Product (GDP), the favorite measures of economists to assess our national progress. Devised by Samuel Kuznets in the twentieth century, GNP and GDP are simply cash-based measures of the total goods and services produced in an economy. During World War II, Britain adopted the GNP as the primary measure of war production in fighting Nazi Germany. In 1945, with peace and the founding of the United Nations, the World Bank and the International Monetary Fund (IMF), the U.N. adopted the GNP and GDP, which became the standard U.N. System of National Accounts (UNSNA). Gradually, they morphed into the global yardstick of success slavishly reported in all media and followed feverishly by legions of policy wonks—whether social progressives, reformers, or reactionary conservatives. Yet, because GNP and GDP are money measures of output of goods and services produced within a nation, neither looks beyond economics to understand genuine value creation within a society. Today, more and more groups—from environmentalists who value nature, to women who perform most of the world's unpaid work, as well as those concerned with social justice—are critiquing GNP/GDP and a host of other economic indicators.

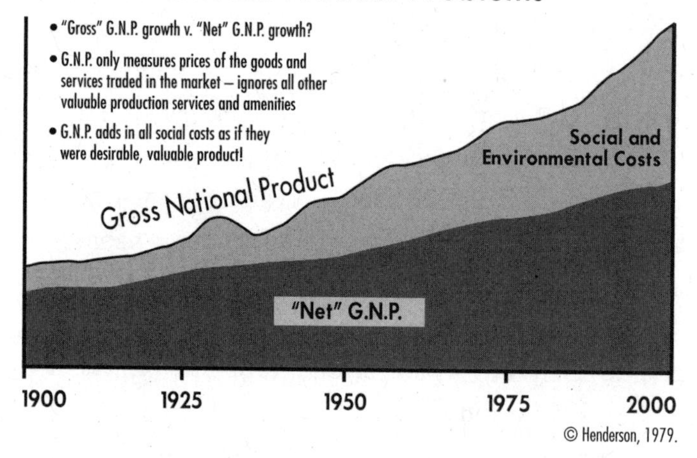

They ask common sense questions that are difficult and embarrassing for economists. Why do they focus on money and factories as capital and so often ignore human and social forms of capital, as well as ecological assets, like rain forests and biodiversity? Why does GNP treat education as a cost instead of a vital investment in our future while not accounting for its quality? Why does GDP go up when people are sick and have to pay for health care but doesn't include ways to measure wellness? Because GDP reports people's average incomes, it obscures poverty gaps (a country's average income can rise when a few millionaires get richer while most people could be homeless). And why does GNP set the intrinsic value of people and the environment at zero? Economists avoid or evade such questions. While the GDP does shed some light on the way we spend our money, it doesn't count many of the things that actually make life worthwhile. In the past few years, these questions have led to debates within the economics profession and to a host of new, multidisciplinary indicators of well-being and quality of life—beyond economics.

So how can each of us also work toward a broader integration of value and values that ensures the health and viability of the places in which we live? Betsy Taylor, who founded the Center for the New American Dream, a nonprofit research group in Washington, D.C., advocates simplifying our lifestyles. "The Center is a group that's working to help Americans live consciously and buy wisely. The Center did a national poll of 1200 adults in August of 2004 and we were asking Americans, 'So how do you feel about the American Dream? What do you think about materialism?' And overwhelmingly people are worn out. One of the most surprising results of the poll was that almost half of Americans say that they have actually taken a pay cut to get more time in their life for things they really care about." (www.newdream.org) Betsy's findings also match others in networks advocating "Voluntary Simplicity," borrowed from the 1973 perennial by Duane Elgin, www.simplelivingamerica.org. Betsy notes some deeper uneasiness, "They're also saying they want

Betsy Taylor
Center for a New American Dream

more time for family, they want more time for community. There are so many costs to our 'more is better' definition of the American Dream. First and foremost is our own quality of life." Betsy thinks the new indicators of success and the new definitions of both economic progress and personal happiness are absolutely critical. Betsy is part of a worldwide movement that surfaced in 1992 at the Earth Summit in Rio de Janeiro to overhaul economic indicators from GNP and GDP to better measures of unemployment, unpaid work, along with social and environmental costs and benefits. One hundred seventy nations pledged at that Summit in its Agenda 21 declaration, to make all these revisions to their own GNP and GDP indexes.

By 1995, the World Bank released a report, *The Wealth of Nations,* which should have led to an overhaul. It found that 20 percent of a nation's capital was financial and "built" capital (financial capital, factories); 60 percent was social and human capital (citizens and social organizations) and 20 percent was ecological capital (nature's life-support systems). Yet this World Bank report was as widely ignored as the earlier Agenda 21. The GNP/GDP habit was still deeply ingrained and reinforced daily by mass media. A more widely accepted set of new indicators is the United Nations Human Development Index (HDI), started in 1990, which ranks nations according to the quality of life and includes criteria such as life expectancy, educational attainment, and the gap between rich and poor and size of military expenditures.

Inge Kaul, a petite, energetic German economist deeply committed to more sustainable forms of human development, was a key pioneer of the HDI, along with the late Mahbub ul Haq, a former Finance Minister of Pakistan. Inge still works with the United Nations Human Development Program as its director of Development Studies. Inge explains, "What this indicator—the Human Development Index—really tells us is how nations translate their economic growth—to the extent that they have it—into human well-being. What we try to do with the HDI index is to say, 'Now we work hard, we undertake all kinds of economic effort but at the end of the days are people better off than they were before?'" She adds, "what you find in the index is life expectancy—the opportunity to live a long and healthy life; education—so that you know what exists in terms of opportunities so you can make an informed choice. Then, with a certain amount of income, we have the means to get properly clothed, housed, and lead a rather secure life. So we used these three indicators and we added them up to produce one index—the HDI. We thought we would succeed in turning around people's optic about what is progress, what is success, what makes for a good life." The HDI has been adopted worldwide as an alternative, broader measure of human development and over forty countries now produce their own domestic versions of HDI. The highest-ranking countries often cite their HDI rank in their tourist information. The United States' ranking is often lower than Canada and the Scandinavian countries, due to our higher infant mortality rates. In the 2004 HDI, the most livable countries were Norway, Sweden, and Australia. In 2005, Norway again topped the list, followed by Iceland, Australia, Luxemburg, Canada, Sweden, Switzerland, Ireland, Belgium, and the United States in tenth place.

Inge Kaul
UN Development Program

Bhutan takes seriously people's right to the pursuit of happiness. A Buddhist nation in Asia, Bhutan may not have economic might but does have a wealth of inner resources. Bhutan has developed Gross National Happiness (GNH) indicators because in its philosophy, happiness should be considered along with money and GDP growth. Many indicators in Bhutan focus on similar trends as the HDI, with an emphasis on preserving Bhutan's peaceful, contemplative culture, while balancing these concerns with outside investment influences and selective economic development. Bhutan's GNH sparked enormous media interest. Even the London-based journal, *The*

Economist ran a three-page story "The Pursuit of Happiness," (Dec. 18, 2004) and *Technology Review* in its January 2005 issue explained in "Technology and Happiness," why more gadgets don't necessarily increase our well-being. All the attention on Bhutan led to a tourist boom that is being carefully controlled. Many economists and social scientists quietly studying human "satisfaction" found themselves in the limelight and happiness surveys are now, happily, fashionable and proliferating. A *New Yorker* review of books on happiness (Feb. 27, 2006) covers psychological studies including Jonathan Haidt's *The Happiness Hypothesis* (2006) and economist Richard Layard's *Happiness: Lessons from a New Science* (2005) but places the launch of positive psychology in the late 1990s rather than in the 1960s by Abraham Maslow in his ground-breaking *Toward A Psychology of Being* (1963). Erasmus University in Rotterdam runs a world database on happiness, while economists tend to agree with Daniel Kahneman, a Bank of Sweden prizewinner, who finds that income levels have little to do with how much people enjoy their lives. He hopes to establish an official measure of national well-being in the United States to complement GDP. This is the purpose of the Calvert-Henderson Quality of Life Indicators I cocreated with the Calvert Group of socially responsible mutual funds, launched in 2000 and updated regularly at www.Calvert-Henderson.com, which we use in our TV series. All the ferment about the goal of "human development" (to which economic growth had been assumed as the means) led to more focus on humanity's life support system: planet Earth, and the sustainability of the biosphere and its vast ecological assets. *TIME* in its January 17, 2005, article reported that where people live is a key influence in "feeling good," based on studies of Subjective Well-Being (SWB) by psychologists Robert Biswas-Diener of Portland State University and Ed Diener of the University of Illinois. Latin Americans in many studies are among the happiest people in the world—among the least happy were Russians, Lithuanians, Japanese, Chinese, and South Koreans. A new Happy Planet Index from the New Economics Foundation in London reached similar conclusions.

The Swiss economist Mathis Wackernagel, developer of the *Ecological Footprint Analysis* sums up his environmental, resource-based approach to human well-being. "There are these two big stories in civilization that seem to be in contradiction. On the one hand that we have better and better lives, that we use more and more resources, more chocolate, more houses, bigger cars. And at the same time, ecological capacity is going down—less trees, more CO_2 in the atmosphere, less water. So we have ecological degradation happening at the same time as we increase our demand for resources." Mathis prepares an annual survey for the World Wildlife Fund, "The Living Planet Index," which monitors all countries to see how their own resource-consumption fits within their nation's natural resources. The results are published in a map showing which countries exceed their resource-capacity and rely on imports from other countries and which can stay within their own resource-base. On a planetary basis, Wackernagel finds that the human family is consuming more resources than the planet Earth can provide on a sustainable basis. Dr. Wackernagel adds, "The Ecological Footprint is a very simple resource accounting tool. On the one hand it measures how much nature we have and on the other hand how much we use. It's interesting that actually some lending institutions have used the Ecological Footprint to evaluate their countries in terms of financial risk. Are they ecological debtors? Does that mean they use more resources than they have available? Or are they creditors? Now the creditors are countries like the green super powers, Brazil for example or Russia." Mathis emphasizes that it doesn't mean that these countries use their resources very wisely, it just means they have a lot compared to what they use. "We have the debtors, those who actually use more, like Switzerland, where I'm from. That's why, for example, we worked with London. The London

Mathis Wackernagel
Ecological Footprint Analysis

Business Council—not just a green group—actually sponsored a study to look at how London can stay competitive using their ecological footprint because as resources get scarce cities that are not positioned for the future will have a harder time" (www.footprint.org).

Using resources and energy more efficiently is the key to sustainable human development as we explore in chapters 4 and 8. The Ecological Footprint can help. The country with the largest footprint is the United Arab Emirates because it uses so much air conditioning. Reducing waste is the key to both quality of life and a healthier economy. For example, "Pulling the Plug on Standby Power" (*The Economist*, Mar. 11, 2006) says that just adding better technology and higher standards to hundreds of U.S. appliances, from TV sets to microwave ovens, wasteful standby features could save 5 percent of total U.S. residential electricity consumption and $3 billion annually for consumers.

Venezuela is also working to reduce its Ecological Footprint. However, the oil industry still poses a significant threat to the environmental health of the nation as described by Frank Bracho, former Ambassador of Venezuela to India, and author of many books on sustainable development, globalization, the role of oil, and the need to shift societies' resource bases to solar and renewables. Frank also edited a groundbreaking study of alternatives to GNP in 1989, *Toward a New Way to Measure Development*, still available in Spanish and English. Frank says, "The relationship of Venezuela with oil to me can be best defined in terms of 'petro-addiction.' That's the way I basically view it because like dependency on drugs the patient knows that it causes harm, yet he keeps on consuming. It's fundamental to redefining success and to redefining wealth and progress, because progress cannot be engaging in some kind of wealth in the immediate term that in the long term is going to prove suicidal." Frank describes his work in Venezuela and internationally, "We have also undertaken activities to promote the redefinition of wealth and progress, not just nationally but internationally, in terms of seeking a substitution for the Gross-Domestic-Product-based wealth paradigm which doesn't really tell you much about the quality of life, or about the sustainability of what you are doing. It only measures in monetary terms what is produced in an economy without telling you whether it is destructive or not, whether it's at the expense of the environment or human health. The most important things in life are not prone to measurement."

Betsy Taylor, Mathis Wackernagel, and Frank Bracho all agree that GNP and GDP must be overhauled and many new, broader indicators must be included. It's not only necessary to overhaul these outdated Systems of National Accounts to include unpaid work (the unmeasured 50 percent of all human production, which we look at in chapter 3) but social and environmental costs and benefits as well. Properly accounting for public infrastructure, paid for by taxpayers (e.g., roads, hospitals, schools) must be recorded as assets, not only as debts. The Calvert-Henderson Quality of Life Indicators have long called for U.S. expenditures on education, research, and development to be recategorized in GDP from "consumption" to "investment" in our most important asset—our children. *Business Week*'s cover story "Unmasking the Economy" (Feb. 13, 2006) made the same points, that the United States is now a knowledge-based economy and knowledge is a key factor of production. Statistics matter! I have urged these changes in many meetings of statisticians in Europe, Japan, China, and Latin and North America for many years. At last, in the largest-ever conference of over seven hundred statisticians in Curitiba, Brazil in 2003, the group endorsed the need for GDP to include asset accounts to record the enormous value of taxpayer funded public infrastructure, to balance GDP's current recording only as debt. This stroke of the pen correction (called "accrual accounting") was instituted in the United States in 1996 and accounted for about one third of the Clinton administration's budget surplus (the rest was from cuts in military spending and high tax revenues from the "dot. com" bubble). Canada followed suit in 1999 and went from a deficit to a $50 billion (Canadian dollars) surplus (Henderson, 1999).

The new indicators are also changing corporate balance sheets. Verna Allee is a management consultant who helps her clients evaluate less tangible measures of success. Verna breaks new ground in measuring "intangibles" (i.e., knowledge, patents, goodwill, etc.) and designing new accounting standards and protocols in Europe and North America in *The Knowledge Evolution* (1997) and *The Future of Knowledge* (2003).

Verna Allee says to her corporate clients, "There's nothing particularly challenging about a financial audit, but if you're trying to develop indicator sets for whether you're being a good citizen or contributing in a sustainable way to the environment, that's very challenging work." Dr. Allee points out that consumers make choices every day about what they buy. "Do I really want to wear that brand on my T-shirt? If a company can tell the story of its success only in financial terms that's a very limited story." Verna cited a 2004 survey by Deloitte Touche that determined that 72 percent of the population felt that they would prefer to work for the company that somehow gives back to social good. "That's a much bigger story about a company's success." Allee's deeper point is that these are two completely different worldviews as well as two completely different sets of assumptions about success and about economics. These worldviews and assumptions led to the new national and global indicators of wealth and progress that are challenging the GNP/GDP view. As companies are forced to internalize environmental and social costs in their balance sheets, the good news—as many have discovered—is large savings and greater profits. Why haven't companies done this sooner? Because faulty economics ignored such longer-term costs and risks, as I have documented elsewhere. A pioneering measure of corporate environmental performance is the "sustainable value" analysis (www.sustainablevalue.com) devised by researchers Frank Figge in Britain and Tobias Hahn in Germany (*Environmental Finance*, June 2006).

Major stock exchanges in New York, London's FTSE, and Brazil's BOVESPA all now feature indexes of companies that are cleaner, greener, and more ethical. To the surprise of economists and financial analysts these socially responsible indexes have outperformed Standard & Poor's 500 and now account for almost $2.3 trillion in assets managed in the United States alone.

Tim Smith is vice president of the Boston-based Walden Asset Management and president of the Social Investment Forum, the trade association of socially responsible investors, mutual funds, brokers, and other asset-managers in the United States. Tim sees the astounding growth of this sector of U.S. capital markets as the emerging mainstream. "We're not talking about esoteric social and environmental issues. We're also talking about issues that have a direct impact on the bottom line and affect the shareholder's interest in a very tangible way. What I'm excited about is watching the number of cities, states, foundations, mutual funds, religious investors, and individuals who are energetically involved in engaging companies now and asking them to change." Tim is a long-time activist investor who founded the Interfaith Center for Corporate Responsibility of the National Council of Churches of Christ in the United States. He emphasizes, "There are investment opportunities that give an investor a chance to say, 'I choose not to invest in America's worst polluters.' In other words a portfolio that reflects your values, that reflects your morals."

This growing debate about deeper values beyond money is now changing the political landscape in the United States. Religion has reentered politics in many ways—from the church-led ad campaign about gas-guzzling SUVs "What would Jesus Drive?" to Dr. Linda Seger's *Jesus Rode a Donkey* pitted against fundamentalist supporters of Republican TV evangelists and the growing "megachurches" catering to lifestyle issues. The National Association of Evangelical's representing thirty million people in forty-five thousand churches is now voicing concerns about climate change and the need for more concern about global poverty and even backs the International Criminal Court opposed by the Bush administration. Meanwhile, the Democrats are trying to modify their former secularism and trying to reclaim the values debate. Canadian opinion researcher Michael Adams has touched a chord in his *American Backlash* (2004), which surveys Americans using a values and lifestyles approach. A massive polling project by the California-based Foundation for Global Awakening found most Americans wanting leaders with higher moral standards and behavior ("The New America," 2004). The new ferment about deeper goals and values has intensified with globalization, outsourcing, and new worries about corporate crime, loss of pension benefits in corporate bankruptcies, cutbacks in health benefits, and whether one's pension plan is invested in "good" companies that can be trusted.

Some employers are beginning to offer options. The federal employee's pension fund, called the Thrift Savings Plan, offers five different investment options but not one of them offered a portfolio based on socially and environmentally responsible standards. The U.S. Environmental Protection Agency (EPA) also offers its employee's the Thrift Savings Plan. It was natural for Brian Swett an employee of EPA to want another option for an environmentally responsible portfolio of companies as his retirement plan. "We're trying to address certain issues because we have causes and beliefs and values that we're trying to support and move forward. To invest your money in a way that is counterproductive to furthering the causes that you believe in, to me, was just hypocritical. I just think it would be fantastic to provide federal employees with an option to invest in a socially and environmentally responsible way that is in line with the work that we do as government employees."

So far, the Thrift Savings Plan still does not offer such a choice of portfolios. However, many public pension plans, following CALPERS lead, are offering socially responsible choices. Many are leading the change in demanding that the companies in their portfolios disclose their plans to help curb CO_2 and other pollutants now confirmed to contribute to climate change and global warming. Tim Smith says, "You know one of the myths that concerned investors face again and again is you're going to lose money. You're investing with your conscience and you're going to get a lesser return. Thank goodness there are decades of history now that show that responsible investors are also prudent investors."

Indeed, as we discuss in later chapters, socially responsible portfolios and indexes regularly outperform traditional investments. Many churches have led the way in social investing. They see the promise of a strong financial return and social impact by investing their pension funds the socially responsible way, as Vidette Bullock Mixon of the General Board of Pension and Health Benefits of the United Methodist Church, explains. "Our investments have consistently been in the top quartile of investment. I would say that we have consistently made a positive fiscal return on behalf of our participants. That is the goal of our pension fund, that we will invest money to bring about a positive return, to influence companies in which we invest to be better corporate citizens." The United Methodist Church Pension Fund is a signatory to the United Nations Principles of Responsible Investing (see www.ethicalmarkets.com). Churches also led in challenging company policies at annual meetings, as we explore in chapter 9. Over six hundred companies worldwide can now account for their success using the integrated triple bottom line: "People, Planet, and Profits." Dr. Judy Henderson an Australian pediatrician, chairs the Global Reporting Initiative (GRI), which sets global standards. Judy Henderson speaks of peoples' personal commitments, as well as her own, "Some of us are shareholders and we can be involved in shareholder activism through that mechanism, but even if we're not shareholders, we are customers, clients, and employees, and we go into the shops and buy products. We like to work for companies that have a good reputation. So in one way or another we are all involved in our daily lives in these processes. It's important for us to be associated with companies that have a good reputation that we can feel good about, buying their products or working for them." You can visit the GRI and watch its progress in signing up new global companies to its "People, Planet, Profit" auditing standards at www.globalreporting.org.

Vidette Bullock Mixon
United Methodist Church

All of the new indicators of sustainability and quality of life, at every level, have forced soul-searching amongst the economics profession—now badly split between traditionalists who pooh-pooh the new indicators and see markets as self-correcting—and the innovators—environmental-economists, social-economists, evolutionary economists, as well as those from many other scientific fields. A new debate about methodologies in the economics profession emerged at the 2004 Nobel Prize awards (see www.hazelhenderson.com, click on "Editorials"). The Economics prize is not a Nobel but was set up by Sweden's Central Bank in 1969. The actual name is the Bank of Sweden Prize in Economic Science in Memory of Alfred Nobel. Some of the scientists who think this prize should be delinked from the Nobel Prize and properly named, include Professor Robert Nadeau, a historian of science at George Mason University, who says "Recently Peter Nobel, who is the head now of the Nobel family, made the allegation that the Nobel Memorial Prize in Economics is essentially infringing on the trademark of the Nobel Prize. What he was saying in effect was that the Bank of Sweden created this prize and this prize is very different from other prizes given by the Nobel Prize committee and that he personally objects to the suggestion that mathematical theories used by mainstream economists, such as those of the University of Chicago school, are scientific." Dr. Nadeau has written his views in his *The Wealth of Nature* (2003), a rigorous critique of the economics profession and why from the view of environmental science, economics is not and can never become a science. Another critic is Ralph Abraham, a mathematician at the University of California in Santa Cruz, who adds, "It must have come to the attention of Mr. Nobel that there is something really fishy with this economics prize which is

Nobel Prize Medal

always awarded to people from this one arcane and very questionable economic theory." Abraham agrees with Nadeau's view that it wasn't Nobel's original intent to create this prize and the prize is really not consistent with his purpose. "The prizes were to be awarded, said Nobel in the original letters, to people who advance human knowledge in service, in the attempt to improve human life, and to enhance human survival." Ralph Abraham adds that, "It looks to me highly probable that the so-called Nobel Prize in Economics will be re-named or will be abolished." Economists usually define and measure success. Now their methods are being questioned. Even *The Economist* (Dec. 24, 2005) admitted that economists, including Herbert Spencer, an early contributor, had actually coined the phrase "survival of the fittest," so often attributed to Charles Darwin. New research on Charles Darwin (www.thedarwinproject.com) sets the record straight. Darwin thought that although competition between species was important in evolution, that the human genius for bonding, cooperating, and sharing was the key to our survival—including the evolution of moral sentiments and altruism. *The Economist* article admits the error in economics was in over-emphasizing competition and underplaying the equal importance of cooperation in humans. Indeed, the evidence of the evolution of human societies from roving bands of nomads to villages, towns, and great cities to multilateral organizations, the European Union, and the United Nations bears out this key role of cooperation. Yet, as I have noted elsewhere, business schools still teach narrow competition and short-term profit-maximization, which many scholars believe fostered the bad behavior of many executives at Enron, WorldCom, Parmalat, and other corporations.

Ralph Abraham
Professor, University of California

We agree with Robert Nadeau that mainstream economists really believe that their theories are scientific. "They are for the most part very idealistic people. They simply don't realize, I think, as a result of their own training that the theories are not scientific." Ralph Abraham declares that "The global economy is suffering enormously from the poor advice of the professional economists who are basing their advice on a completely unproved theory." Robert Nadeau adds: "The managers of the World Bank and the IMF and the regional banks receive advice from economists which they believe to be based on a scientific paradigm, the economists then make prescriptions for dealing with economies in the developing nations and the underdeveloped nations and they have inflicted enormous harm on millions if not billions of people." Ralph Abraham sums up the problem saying, "Metrics of economic importance that respect the environment, women's work, and so on, will be just part of an overhaul of the economics profession. So now is the time, I think, to bring these new metrics and environmental ways of thinking into the arena of public discussion."

Robert Nadeau
Professor, George Mason University

So, in spite of this wide debate and all these new ways of measuring genuine human development and quality of life, how was it that Wall Street and our financial media missed this story? Why are we still using old economic scorecards—like GNP? Why do most of our economic textbooks, computer models, and business schools still assume that more money is the only bottom line? Not only the seven hundred innovative statisticians who convened in Brazil at the First International Conference on Implementing Indicators of Sustainability and Quality of Life in 2003 but many others are at work, from the cities of

Shanghai and Sao Paulo to Canada's new Measure of National Well-being. The Calvert-Henderson Quality of Life Indicators, which we use on our TV series measure twelve broad aspects of quality of life in the United States, including employment, income, education, health, human rights, infrastructure, public safety, national security, shelter, environment, energy, and recreation and culture. So, why do the old methods persist? Force of habit, together with old worldviews entrenched in business, government, academia, and labor unions reinforced by mass media are the likely causes. Economists still call those hidden social and environmental losses "externalities"—costs that companies externalize from their balance sheets and pass on to society or future generations. Calvert-Henderson Indicators measure these hidden costs and the hidden wealth in our communities.

Wall Street, too often forcing companies to show ever-increasing growth every quarter—and sometimes fudging their figures, will punish a slowdown by a negative security analyst's report—which can knock down the stock price.

The GNP-growth scorecard still drives Wall Street, too often forcing companies to show ever-increasing growth every quarter—and sometimes fudging their figures. A slowdown will be punished by a negative security analyst's report—which can knock down the stock price. In the 1990s deregulation globalized financial markets, speeding up investment flows and spreading the growth mania to all stock exchanges. Outsourcing production and jobs to low wage countries improves corporate profits. Yet, cutting costs to compete with Chinese prices is incompatible with even the lowest U.S. wages. As U.S. producers shut down, China has become the world's premier manufacturing country even though wages are now rising there and companies are seeking cheaper workers in Vietnam and Cambodia. All these issues interact. Replacing employees with machines and energy is rewarded in our U.S. tax code. At current oil prices, the United States, still dependent on foreign supplies, has racked up trade deficits of 7 percent of our GNP-measured economy. Why not tax waste, resource depletion, and pollution? *New York Times* columnist Thomas

Friedman advocates such taxes and a "Manhattan Project" style shift to solar and renewable technologies. These technologies have been waiting in the wings since the 1980s as I documented in *The Politics of the Solar Age* back in 1981. Only perverse subsidies to powerful oil, coal, nuclear, and fossil-fueled industries have hampered their introduction—and continue under the U.S. Energy Act of 2005. Shifting taxes from incomes, payrolls, and jobs to waste and pollution, and ending these wasteful subsidies would, at last, allow solar and other renewables to compete on a level playing field. Countries in Europe, Asia, and Latin America are actively debating all these tax shifts, ending perverse subsidies and adopting new scorecards.

We can foster innovation by rebalancing old productivity measurements among labor, capital, and nature's resources, as I show in "The Politics of Productivity Measures" (www.Calvert-Henderson.com, click on "Current Issues"). When nature's contributions to production are properly accounted, we conserve resources, design and operate our industries more efficiently, and save money—as so many eco-efficiency studies show, as summarized in *Natural Capitalism*. Too often investors become day traders and speculators betting on higher stock prices. Such short-run profits often conceal longer-term liabilities, whether environmental or social costs lurk off the balance sheet. Fortunately, many new indicators are more faithful to the truth. While Calvert-Henderson Indicators and the many others we have discussed expand our view of real wealth and progress beyond GNP, we also see how better accounting methods and full cost prices are guiding companies, investors, and consumers to better decisions. As statisticians worldwide continue redefining true prosperity—real human development—using new measurements, recalculating GNP to add human, social, and ecological capital, and subtract pollution—we can see more clearly what success means in our own lives.

ROUNDTABLE
WALKING THE TALK

Ray Anderson with Simran Sethi

"Walking the Talk" is the real litmus test of success for companies. Ethical Markets demands transparency and requires that any company claiming its superior social, environmental, and ethical performance be carefully monitored. Thus, Ethical Markets chose a group of the best-qualified corporate auditing firms that specialize in these aspects of corporate performance. Our "stakeholder analysts" volunteered to interview CEOs of such companies that publish higher codes of conduct and hew to such superior standards.

Interface Carpet, the world's largest manufacturer of industrial carpeting, based in Atlanta, Georgia, is a benchmark company. CEO Ray Anderson is a preeminent leader in steering his company toward sustainability. Our stakeholder analyst, Hewson Baltzell, President of Innovest Strategic Value Advisors, the preeminent global ethical audit firm, interviews Ray Anderson, with host of the show, Simran Sethi.

Hewson Baltzell

Hewson Baltzell: For over ten years Interface has been a leader in sustainability and this is in a sector that is difficult, it's a high environmental impact sector, basically, oil-based making of carpets. Interface has not only made useful energy efficiency efforts but has gone much farther than that by looking at a full "cradle to grave" approach designed for the environment—looking at where the materials come from, how they're processed, and then how those materials end up eventually at the end of their life—back into the landfill, etc. Interface stock has been trading around $13 up from $3 a share back in mid-2003 so it's a big increase although it's substantially down from its high of over $20 back in the late 1990s. Is there any relationship—good or bad—between Interface's stock price and your efforts in sustainability?

Ray Anderson: We might not be here today but for the sustainability initiatives that we undertook, the cost reduction efforts, the product development efforts from the point of view of sustainability and the goodwill of the marketplace. When your customers would rather do business with you than a competitor because they believe in what you're doing, this has a galvanizing effect on our people. So all those things working together on a highly leveraged company—over $450 million in debt—we've been able to survive this huge downturn in the marketplace and thrive and gain market share in the process. And as the economy in our sector recovers we stand to set a very good example of doing well by doing good.

Simran: What social initiatives helped Interface move toward sustainability?

Ray: It begins with safety in the workplace, a prime focus. That's true for any company and it's where you really must begin. We learned the hard way that social equity begins at home, too, as we made a number of acquisitions through the years we were putting together distribution channels though acquiring carpet dealers and contractors around the country. We were integrating all of these new companies into Interface and we found that cultural sensitivity is very important and we weren't very good at it. We didn't do a good job of bringing all those disparate groups together and melding them into a unified company. It cost us dearly so we found that social equity really begins with how you treat your own people. We've tried to be good corporate citizens in the areas where we operate. We've got recycling going where it didn't exist before in this city or that city. So much of what we're doing even on the social equity side is driven from the environmental side. The environmental ethic permeates most of what we do.

TWO

Global Corporate Citizenship

THE GLOBALIZATION OF FINANCE AND TECHNOLOGY and the growing power of corporations have transformed the way a company functions in the world economy. According to the Global Policy Forum, over half of the largest one hundred economies in the world are global corporations. Wal-Mart's economy is bigger than 161 countries, including Israel, Poland, and Greece. Toyota is bigger than South Africa, and Mitsubishi is larger than the fourth most populous country in the world—Indonesia. These global corporations impact the lives of millions of people in many countries, as evidenced by rising public concern for corporate accountability for human rights, workplace safety, decent wages, and protecting the environment. History has seen several waves of globalization as explorers from China in the fourteenth century peacefully visited other Asian countries. Later, Europeans violently colonized Africa and the Americas. Current globalization of technologies and deregulated markets, privatization, free trade, and 24/7 currency exchange accelerated processes of global change, as I described in *Beyond Globalization* (1999). The new unregulated "global casino" of financial markets outpaced national politics, sovereignty, and international law. Efforts to create global standards for workplaces, human rights, and environmental stewardship still lag by decades.

Dr. Michael Dorsey, Professor of Environmental Studies, Dartmouth College, sums up these issues: "Overwhelmingly the number one leading type of corporate crime is against the environment. We can work on two fronts in order to move forward on corporate accountability. The first front is working with corporate social responsibility, working with strategies of shareholder activism. This is bringing certain ethical and moral demands to compel companies to be good citizens, to obey the law. The other thing that we really need to do is focus on corporate crime—this has been happening a lot in the United States; New York's Eliot Spitzer has really been pushing this issue! When we talk about corporate crime we can define it in terms of simply companies pleading guilty to charges that were brought against them, and/or paying fines that were levied against them." Auditing firms are also implicated with four global accounting companies now dominant.

Alexis de Tocqueville warned that the fledgling United States contained the seeds of "a manufacturing aristocracy." As companies consolidated into the huge "trusts," seeking monopolies, they began dominating Congress with their economic power.

The recent corporate crime wave in the United States and Europe has deep systemic roots. Alexis de Tocqueville in his famous study of the fledgling United States, *Democracy in America* (1835), warned that it contained the seeds of "a manufacturing aristocracy." Corporate charters were granted by states and were played off against each other by companies demanding ever less regulation and oversight. As companies consolidated into the huge "trusts," seeking monopolies, they began dominating Congress with their economic power. By 1886 the Supreme Court had ruled that corporations were as natural persons enjoying all of the rights of citizens—with few of the responsibilities. As mentioned, business schools, too, are implicated, since their curricula reinforce these views and teach orthodox economics, which allows companies to externalize social and environmental costs from their balance sheets. Within this current legal framework, corporate leaders are required to maximize shareholders return or they may find themselves out of a job. Yet, this maximizing rule is changing toward including other stakeholders, and into longer time horizons in the now-global context, as we discuss in chapter 9. Many corporations are still fighting rearguard actions, as the coal and oil industry did in denying global warming. Yet civic groups keep up the pressure with media campaigns, monitoring companies' social and environmental impacts on hundreds of Internet sites and blogs—calling for rechartering of companies and ending their limited liability protection.

The financial trouble of large companies not only impacts their shareholders in the countries in which they are based but also employees with 401K plans that are invested in these corporations or through their pension funds, whom we call involuntary investors. Labor unions are becoming more active internationally on many issues, including child labor—and increasingly exert pressure on companies they own in their pension plans. Rich Ferlauto, the director of pension and benefit policy in the United States' largest public service employee union is working to ensure that Enron-like fiascoes don't wipe out the savings that people have worked so hard to create. Rich Ferlauto, Director of Investment Policy for AFSCME, the American Federation of State, County, and Municipal Employees, says, "We've got 1.5 million members who are public employees. They work for state and local governments, as librarians, clerks in municipal offices and environmental departments

Rich Ferlauto
AFSCME

supplying services to the public. Their retirement is invested through local and state public pension systems across the country. If we're able to actively tap not only the $2.7 trillion in the public pension system, but the $6 trillion in workers capital if you will—all the retirement assets of workers who are organized in some way—you could have a huge impact on the markets." Rich adds that, "The funds of our members, essentially own everything—they're universal owners. Any benefit that one company would get from global warming, for example, would have an impact on agriculture; polluting on the coast of California will destroy the tourist industry. They're all linked together. Right now, the largest pool of investment capital is in institutional investors and pension funds. So for the past twenty years I've been engaged in a whole variety of ways to figure out how pension capital can be used not only to produce retirement earnings but can also produce social and economic good."

Alisa Gravitz
Co-op America

A new debate about the responsibilities of corporations has already shifted the ground beyond the former view that management's only duty is to maximize returns to shareowners. *The Economist* published its orthodox view of "The Good Company" (Jan. 22, 2005), reiterating that business should stick to maximizing returns to shareholders. The *Wall Street Journal* (Dec. 2005) joined in with its seventeen-page debate "Corporate Social Concerns: Are They Good Citizenship or a Rip-Off for Investors?" between representatives of General Electric, the Competitive Enterprise Institute, and the civic group Rainforest Action Network. Both of these market-oriented publications received similar feedback from their readers. *The Economist*'s surveys of CEOs found over two-thirds believed that good corporate citizenship was good for business, while *Wall Street Journal* readers in a follow-up survey on January 9, 2006, found 77 percent saying social concerns were somewhat or very important to companies. Not only are the economic diehards in retreat but many companies embracing corporate social responsibility in the global 100 Most Sustainable Corporations (from a pool of 2000 rated by Innovest Strategic Value Advisors) are also endorsing the Millennium Development Goals of the United Nations in 2000 to reduce poverty, eradicate hunger, and improve health and education. Alisa Gravitz, Executive Director of Co-op America, based in Washington, D.C., publishes the annual *Green Pages* directory of cleaner, greener, more socially responsible companies. Alisa also helps direct the Social Investment Forum. Alisa says, "If you own shares of a company you know all those proxy statements you get in the mail. Look at them closely because many of them have what we call social and environmental and corporate governance shareholder resolutions on the ballot that you have the right to vote for. If, for example, a company like Exxon Mobil asks in its proxy, 'Do you believe this company should reduce its climate change pollution and invest in renewable energy?' You want to say, 'Yes, I do!' And you want to vote for that change for the company. Another way you can get involved is by investing in socially responsible mutual funds because one of the things socially responsible mutual funds do for you is they are active shareholders on your behalf. And you can find out about which mutual funds are socially responsible at our Web site www.socialinvest.org. And what these companies, the socially responsible mutual funds, do for you is they meet with every company in their portfolio and they ask the tough questions, they demand that the companies continuously improve their socially and environmentally responsible performance."

Dr. Simon Zadek, chief executive of the British NGO AccountAbility, the preeminent international association for promoting accountability across business, government, and civil society is a global leader in corporate social responsibility. Dr. Zadek talks about what it means for corporations to be held accountable. "Just a few of the brands that most of us know and love or sometimes love to hate—the Nikes, the GAPs, the Liz Claibornes, the Sarah Lees and so on, whilst it's always possible to find problems in any large, sprawling supply chain, if one's just looking at the question, 'Have people's lives improved?' I think the answer in many instances would be *Yes*. Nobody would've imagined back in 1988 that fifteen to twenty years on, really every branded company in textiles and garments and footwear would have more or less ILO branded codes of conduct dealing with hundreds of thousands of workers all over the world. So whether you look at these standards or human rights in oil and energy, issues to do with drug pricing, or more recently, obesity, the way in which the financial sector is gradually being pushed to reassess the basis on which they invest, it's really been an extraordinary fifteen years." Yet, this progress must be vastly expanded, monitored, and regulated by new international agreements, treaties, and a new global financial architecture to oversee unregulated capital markets now destabilizing whole countries, as I described in *Beyond Globalization* (1999).

Socially responsible investing got its momentum in an international setting and gathered momentum when Elizabeth Dowdswell, Canadian former chief of the United Nations Environment Program (UNEP), launched her UNEP Finance Initiative in the 1990s. Dr. Dowdswell patiently convened leaders in banking, finance, and asset-managers worldwide, helping them understand the new financial risks and benefits to them and the companies in their portfolios. This pioneer program provided key leverage as financiers began pushing the companies they owned into taking more environmental responsibility (www.unepfi.org). The World Business Council for Sustainable Development, launched at the Earth Summit in 1992, also influenced its member companies with studies of how eco-efficiency in their use of energy and production methods could increase their profitability. This lesson is now commonplace, as we report throughout this book. The newest corporate convert is General Electric, whose new president, Jeffrey Immelt, launched GE's Ecomagination initiative, the brainchild of Lorraine Bolsinger, staff engineer and head of Ecomagination, and managed by Beth Comstock, head of marketing. Even Wal-Mart, trying to salvage both its battered image and its sagging balance sheet, launched its own Sustainable Wal-Mart makeover. Voluminous reports now document thousands of these corporate environmental success stories, including *Natural Capitalism*, coauthored by Hunter Lovins, Amory Lovins, and Paul Hawken. All this led the U.N.'s secretary general to launch the U.N. Global Compact in 2000, which asks companies to engage with ten principles of good corporate citizenship. At first, the Compact did not require or even monitor compliance on the part of companies that signed up until I introduced them to the Calvert Group, which now volunteers its social auditing research. However, civic watchdog groups, including Corpwatch.org and GlobalExchange.org quickly and effectively lobbied for change. Companies are now monitored and those remaining recalcitrant are removed from the Compact. As with all massive paradigm shifts, anomalies persist. The United Nation's own pension fund often ignores the Compact's principles when investing its $29 billion of assets. Many companies in its portfolio, such as Exxon, Rio Tinto, Anglo-American, and other mining companies, as well as Archer Daniels Midland and Wal-Mart are shunned by most socially responsible funds, as Bloomberg reported (Oct. 31, 2005).

Jane Nelson
Coauthor, *Profits with Principles*

Jane Nelson, coauthor of *Profits with Principles* (2004), an associate of Britain's Prince of Wales' Business Leaders Forum, and a former banker, is still hopeful, "I think the U.N. creates a lot of the enabling frameworks within which business happens on a global basis and makes it possible for companies to operate globally. I think the second critical role that the U.N. plays—it sets norms on a global basis for governments. The U.N. Global Compact's ten principles from already agreed U.N. norms and conventions in labor, human rights, environment, and anti-corruption are applied to the business sector. Any business can adopt these ten principles in their own spheres of influence and they're principles that have been agreed by pretty much all the member states of the U.N., all the governments of the world." Georg Kell, executive head of the Global Compact explains further, "The Compact is based on dialogue, learning and partnership projects and it is through the power of good practices that we hope to establish the business case for the principles of the Compact. And financial analysts everywhere increasingly give a premium on pro-active principle-based approaches. Once financial markets share the perception that it makes sense to embrace these principles then I think we have achieved our goal. We are talking here really of universal principles whose implementation at the end of the day needs to be vetted by society at large." Thus, Kell defends the U.N. Pension Fund by admitting that the U.N. can never enforce implementation "even if we had a thousand analysts at our disposal." Debra Dunn, former senior vice president of Hewlett Packard, an early signatory, adds, "I think we have unique opportunities through the Compact to make important progress in areas that will determine what kind of place this planet is to live on in the next ten to twenty years."

Georg Kell
U.N. Global Compact

Jane Nelson pinpoints some key issues: "I think so much of the debate on responsible business tends to divide into two questions:

Is it regulation? Or is it voluntary? Both have their pros and cons, and both have their role. I think we do need to raise the base of certain regulations both in this country and, certainly, on an international basis. But I think governments have an enormous amount of other levers at their disposal, which we're probably not thinking about creatively enough. The British government passed a regulation for pension funds requiring all pension trustees to say whether they take social and environmental issues into account when they're selecting their fund managers and stock selection processes. They're not saying you have to take it into account. They're just saying you have to disclose whether you do or not. And just that procedure has got pension fund trustees at least thinking about the issues and having conversations with their fund managers—you know, 'How are we addressing the issues,' etcetera?"

> I think so much of the debate on responsible business tends to divide into two questions: Is it regulation? Or is it voluntary? I think we do need to raise the base of certain regulations both in this country and, certainly, on an international basis.

Indeed, this U.K. rule has become part of the European Union's requirements and has shifted billions of dollars worth of assets into socially screened portfolios, as covered in the Triple Bottom Line conferences held by Amsterdam-based BrooklynBridge.org. Today, pension funds representing over $3 trillion are major activists on climate change, publicly announcing that if the companies in their portfolios don't disclose their climate risks and plans to mitigate them, they may be dropped from these pension plans. The big reinsurer, Swiss Re, started shifting its portfolio from fossil fuels to renewable energy companies in 1993.

Innovest Strategic Value Advisors (www.innovestgroup.com) published a report that documents how corporate responsibility and socially responsible investing can improve a company's performance. Hewson Baltzell, cofounder of Innovest and an Ethical Markets stakeholder analyst, explains that with Innovest's new analyses of social, environmental, and ethical risks corporations now face, companies that mitigate these risks actually have improved their financial performance. "At Innovest we rate companies based on sustainability issues, environmental and social issues, and we apply those ratings to portfolios. On behalf of a public California Pension Fund we conducted a simulation using their own portfolio, that is to say the portfolios of their own asset managers. In this report we look at five of those managers. The managers have different styles—large cap, growth, large cap corp holdings, global, etc. This simulation started in 2002 and went through 2004 using the actual portfolios that were created by those managers at the beginning of every quarter. We take that portfolio, apply our ratings, and we buy more of the companies that have high Innovest ratings and less of companies that have low Innovest ratings, so essentially over weighting the ones that we think are the high performers. The results over the three years in each of those five portfolios are that the addition of our information has provided value. That is to say that our information has actually resulted in higher performance, higher investment returns for those portfolios."

Promoting earth ethics is often a good business decision. A corporation's profitability depends on its relationships with local and global communities. Mallen Baker, of Business in the Community, a U.K. organization, explains, "Business impacts communities in all sorts of ways, and the most obvious of those is that it creates jobs. Employees work and live within those communities, and some of that wealth hopefully goes into the communities. Mostly what businesses are seeking to do when they invest in a community is to improve the health of the community, improve the education facilities, the local health facilities, and so on. There are a number of reasons why they do that, but principally, it is the fact that their employees, whom the companies are very keen to hold on to, are living in the community. They will stay with the company if they see that the company cares about the community, and secondly, that community is one where they want to bring up their kids. Business in the Community, a movement of 750 businesses, committed to corporate social responsibility twenty-two years ago when there were inner city riots in the United Kingdom. A group of senior business leaders got together and realized they needed to invest in the health of the community where they operated for their businesses to be equally healthy. Business in the Community at that stage focused on getting business resources into communities' economic regeneration. Since then the group has broadened out to look at all the different aspects of corporate social responsibility." Britain has innovated the social enterprise in a new law to foster companies with social and environmental missions (*The Economist*, Nov. 26, 2005).

Mallen Baker
Business in the Community, U.K.

The accounting profession has converted its methods of analysis using new integrated triple bottom line parameters through the Global Reporting Initiative, as GRI Chair Dr. Judy Henderson (chapter 1) elaborates, "The Global Reporting Initiative is now recognized as the framework for sustainability reporting at an international level. Ten years ago corporate governance's relationship with long-term value creation was just not in the picture. Now, after Enron and WorldCom, nobody disputes the fact that good corporate governance is related to long-term shareholder value. I believe that in the future the sustainability performance will be in exactly the same position and no one will dispute the fact that good sustainability performance by companies is related to the long-term value creation for their shareholders and their community."

Even the World Economic Forum, which regularly convenes corporate chiefs and heads of state in Davos, Switzerland, joined the corporate responsibility movement in 2005, by promoting the "World's 100 Most Sustainable Companies" (www.global100.org) mentioned earlier. This list still provokes skepticism amongst civic watchdog groups. Meanwhile, Brazil, the world's tenth largest economy, has its own pioneer organization, founded by Oded Grajew—the Brazilian entrepreneur and founder of the Ethos Institute for Business and Social Responsibility on whose international Advisory Board I serve (www.ethos.org.br). Oded originated the World Social Forum, created to counterbalance the corporate globalization model developed at the World Economic Forum in Davos, Switzerland. He explains, "The mission of Ethos is to mobilize and civilize companies to implement socially responsible management in the way to make them partners to build a sustainable and just society. This is a non-profit organization. After six years we created Ethos—about 940 companies representing more or less 33 percent of Brazilian GDP. We must change. Our future will be decided in these next years—our future and certainly that of our children. We work to create an environment that pushes or helps companies to be more socially responsible. We work with the media to be supportive of good practice and to be critical to companies that are not socially responsible. We work with consumers and we create another NGO in Brazil—AKATU—to promote socially responsible consumption. We work with universities and try to implement social responsibility in the curriculum."

Oded Grajew
Ethos Institute, Brazil

Education now plays a larger role in the development of sustainable business practices. And increasingly, top business schools are offering courses in corporate, social, and environmental responsibility, including Amana-Key Desinvolvimento & Educacao, Brazil; Case Western University's Center for Business As Agent of World Benefit;

Presidio School of Management, MBA in Sustainable Business; Sustainable Global Enterprise Program, Cornell University; Base of the Pyramid Learning Laboratory, University of North Carolina; The Corporate Environmental Management Program, University of Michigan; Boston College Center for Corporate Citizenship; Bentley College Center for Business Ethics; the Bainbridge Institute in the United States and others in Europe, Japan, and China.

Jane Nelson also directs a Corporate Social Responsibility Initiative at the Kennedy School of Government at Harvard—rather than the Harvard Business School, which is still conservative on these issues. Jane says, "What we're planning to do is to set up a program which is going to be working with students, with companies, with faculty, academics, really to start a new conversation about the role of business in society and the public role that the private sector plays. So the program's based at the School of Government but working with various centers around the school—Center for Public Leadership, the Harvard Center for Non-Profits, the Center for Business and Government, and the Schoenstein Center for Press, Politics, and Public Policy, which we feel has a very important role to play in this whole field. Today, more chief executives say that young people ask key questions. What's the company's corporate responsibility policy? What is the company doing on these social and environmental issues? As we'll see more and more of the best talent coming from the best universities around the world are going to be asking those questions when they're interviewed for jobs." Student groups, including NetImpact.org and Brussels-based AISEC.org, belong to a new generation that cares about these issues and demands that corporations change their ways.

Today, more chief executives say that young people ask key questions. What's the company's corporate responsibility policy? What is the company doing on these social and environmental issues?

So, evidence abounds that corporations can be good citizens—and provide good financial returns. Hundreds of studies now show that companies do well by doing good. Most agree it's the hallmark of good management. Stock indexes of socially responsible companies out perform the Dow Jones and Standard and Poor's indexes. *The Economist* and other diehards still ridicule corporate social responsibility as mere public relations. Strangely, they side with critics of capitalism itself. They are on the wrong side of history. Companies that maximize profits for shareholders often harm others and the environment, and short-term profit-taking often reduces long-term returns. Many trends in our new century are reinforcing today's shift toward better corporate citizenship, but better global standards and enforcement at all levels are still necessary.

Global communications and 24/7 financial markets birthed the new global superpower: world public opinion. Billion dollar brand names and stock prices can be devalued in real time. Millions of concerned investors, employees, consumers, and faith-based humanitarian groups linked worldwide challenge business and government leaders to serve the public interest. They propose fairer work and trade rules; protection of human rights and the environment. Democracy is spreading around the world. In spite of the failures in Iraq, Freedom House now defines eighty-six countries as free democracies. WorldAudit.org places the United States at fifteenth in their democracy rankings—behind Finland, Denmark, New Zealand, Sweden, Switzerland, Norway, Netherlands, Australia, the United Kingdom, Canada, Austria, Germany, Belgium, and Ireland. Governments and corporations are responding. Leaders competed to donate to Asia's tsunami victims. Many are facing today's global problems and opportunities cooperatively—to provide health care, environmental restoration, humanitarian aid, and peace-keeping support. For example, the Los Angeles and San Francisco Chambers of Commerce collected 600,000 signatures for Proposition 82 in 2006, to raise $2.4 billion a year from California's richest 1 percent to fund pre-schooling for all children (*Business Week*, Mar. 27,

Corporations can be good citizens—and provide good financial returns. Hundreds of studies now show that companies do well by doing good.

2006). Sadly, this proposition was rejected by voters. Yet, philanthropy is rising to new levels, led by Bill and Melinda Gates' fortune from Microsoft, Warren Buffet of Berkshire Hathaway, Ted Turner, founder of CNN, and many more encouraged by former president Bill Clinton's Global Initiative. The United States also signed on to the Millennium Development Goals; to reduce poverty and promote education and a group of Brazilian companies now promote them with an advertising campaign "8 ways to change the world!" Over three thousand corporations have now signed the Global Compact on good corporate citizenship, while one thousand now use the Global Report Initiative Accounting Standards. Companies now targeting these goals stand to prosper and create millions of jobs, helping build a global economy that works for everyone.

ROUNDTABLE
WALKING THE TALK

Simran Sethi

The greatest challenge to corporations is that of global poverty and the widening rich-poor gaps within and between countries. The goal is how to serve those two billion humans at the bottom of the income pyramid still earning less than $2 per day. Simran Sethi, hosted Alex Counts, president of Grameen Foundation U.S.A., along with Alice Tepper-Marlin, Ethical Markets stakeholder analyst and president of Social Accountability International, to find out more about how micro-finance and micro-credit empower the poor.

Alex Counts

Alice Tepper-Marlin

Alex Counts: What Grameen Foundation, U.S.A. does is relieves a major constraint that the poor all face. If you look in the Third World countries most of the poor don't have jobs so they have to create their own jobs by running small businesses. They don't even look like businesses like we'd think of them, but it's raising a few chickens, and selling them in the weekly market. These businesses are terribly under-capitalized, they're inefficient . . . people have to work day and night to earn a few pennies. So by giving them a loan on reasonable terms—the same terms a businessman would have—and in giving it in amounts they can digest like $50 or $100, they can suddenly earn more profit and make a step out of poverty. In Bangladesh, this strategy created a whole financial services industry to serve the poor. Hundreds of thousands of people have gotten out of poverty and what we're trying to do is to spread that to other countries around the world.

Alice Tepper-Marlin: What kind of difference does it make to the borrower? In a typical country, if they went to a bank what would they have to pay and then what do they have to pay if they borrow it from one of these micro-credit programs?

Alex: I lived in Bangladesh for six years—in a single village for two years. What you find is that if the poor go to the bank, they're almost always turned away. People assume they're going there to beg or to bother people so they shoo them away almost. If they can get in they maybe pay 15 percent in interest, which is kind of the commercial rate in Bangladesh but another 20 percent in a bribe to get the loan, so it gets very high. They borrow from a loan shark, the loan shark's very efficient—the loan shark will get the money there but the loan shark will want sometimes as much as 10 percent per week interest. Micro-credit programs basically try to lend with the efficiency of the loan shark but actually at the rate that the banks should be lending at—the commercial rate, 15 percent. It depends on the country and what the rate of inflation is.

Alice: What rates do they repay and how does that compare to, say, the experience of a bank?

Alex: Consumer loans in this country, default is 7–10 percent. If you look at the agricultural lending in Bangladesh default is 60 percent. The poor as lazy is a myth. But the poor as hardworking but not retaining the benefits of the work, that's the reality. The experience of the poor, if you put them in a system where they're actually supported by their peers and held accountable by other women in the village through the micro-lending system in its many variations, default rates are almost entirely below 4 percent. We're trying to raise money to give the capital that these programs need to lend to the poor. It's amazing with $100, $125 a family can change their lives, so we're trying to raise those from people in this very privileged country, which then can fund those loans, and then a loan the next year that's a bit larger. We're also trying to attract volunteers—people that have skills, banking skills, technology skills, management skills, and deploy them as volunteers to help build the capacity of these programs. For example, we got a retired guy from Verizon who was looking for meaning, spoke Spanish and we sent him to a program in Chiapas, Mexico, the poorest state in Mexico, and he was able to help them reorganize their operations, become more efficient and actually lend to the poor at a lower interest rate.

Simran Sethi: What are some areas for improvement for Grameen Foundation?

Alex: We're developing a scorecard that measures how fast are people coming out of poverty in a micro-credit program. We need to try to bundle services—health and education services with the loans so people can also get out of poverty faster. Health problems are prevalent among the borrowers and if you are giving them loans, you can also give them other things. We want to be more accountable to our supporters.

THREE

The Unpaid "Love" Economy

Money has proved to be a key human invention, allowing wider, more complex transactions than barter systems can accommodate. Money has morphed over the centuries from cowrie shells, stone tablets, metal (including silver and gold coins), and paper into pure information: blips on hundreds of thousands of computer trading screens in today's global casino, where $1.5 trillion worth of currencies travel around the planet every day. Money serves as a unit of exchange and can be a store of value, unless manipulated or ravaged by inflation. Economists view the world through the lens of money. Everything has its price and national statistics, GNP, investment, productivity—all follow the money. Yet, as noted earlier, at least 50 percent of all production, goods, and services in all industrial societies, and up to 65 to 70 percent in many developing countries, never count in official GNP statistics, because they are unpaid. These non-money sectors that support the financial economy are known as "gift economies" and scholars from many disciplines study them. *The Gift* by author and poet Lewis Hyde, first published in 1979, views gift-giving as a sacred aspect of human community. Linguist Genevieve Vaughan agrees in her *For-Giving: A Feminist Criticism of Exchange* (1997), focusing on how patriarchal societies devalue caring, sharing, and nurturing work, which is usually unpaid and relegated to women. Marilyn Waring, a sheep farmer and New Zealand's youngest member of parliament, explored the ways economics discounted or ignored women's work in her *If Women Counted* (1989), showing how GNP and GDP national accounts in all countries that use these indices excluded such work by designating such family-based services as well as community volunteering as noneconomic. Ann Crittenden's *The Price of Motherhood* (2001) documents all the disincentives and costs women bear. Even mainstream economists are catching on. The Levy Economics Institute of Bard College reported in January 2006 on its conference "Unpaid Work and the Economy," looking at gender, poverty, and the Millennium Development Goals (www.levy.org).

Professor Karl Polanyi who taught at Columbia University was one of the few economists who studied gift economies and their often sacred nature in his essays *Primitive, Archaic and Modern Economies* (1968) edited by George Dalton. Economist Kenneth Boulding, former president of the American Economics Association also reminded us of the moral underpinnings of economics—always deeply embedded in human relationships. In his *Beyond Economics* (1968) he described three basic ways that humans interact: threats, exchanges, or gifts. Mainstream economics claims to be value-free—even a science. Now that money systems are manipulated and financial players dominate agriculture, manufacturing, and public sectors of nations worldwide, we are reminded that unpaid goods and services are just as vital to the health and well-being of individuals, communities, and nations. As individuals, we realize that our time is just as valuable as money, leading to the growth in North America, Europe, and Japan of what I have termed attention economies (Henderson, 1996) as well as the

Economist Kenneth Boulding, reminded us of the moral underpinnings of economics—always deeply embedded in human relationships. He described three basic ways that humans interact: threats, exchanges, or gifts.

best-seller status of *Your Money or Your Life* by Vicki Robin and Joe Dominguez (1992). Many economists and other social scientists have worried that as more women entered the workforce, family life and children would suffer and birthrates would fall. In those developed countries, including the United States, *The Economist* found the opposite to be true, with higher fertility rates than in Germany, Italy, and Japan, where fewer women work (Apr. 15, 2006). Children tend to be cared for by paid caregivers and housework performed by paid cleaners adding to total employment.

Most people aren't aware that financial analyses and economic policies at state, local, and national levels are money-based statistics that reflect only half of the full range of production, services, investments, and exchanges in societies. The equally important non-money sectors in many countries that constitute, in reality, the core wealth of societies include raising children, the care of families and communities, barter clubs, and cooperatives. They include volunteer activities. (In the United States approximately 89 million Americans participate in volunteer activities at least five hours a week.) When money systems go into crises, such as in Argentina's default in 2001, or after the U.S. Great Depression, when thousands of banks collapsed, people remember that they can create their own local "scrip" currencies, barter clubs, flea markets, and community credit-systems to keep local exchange and production humming. Pictures of thousands of such local currencies appear in *Depression Scrip of the United States, Canada, and Mexico,* lovingly complied by Ralph A. Mitchell and Neil Shafer and self-published by Krause Publications, Iola, Wisconsin (1984).

Scott Burns, financial editor of the *Dallas Morning News* and the author of *The Household Economy* (1977), focuses on the average American family and the many ways it is a lot more productive than the Internal Revenue Service recognizes. "The reason I wrote *The Household Economy* was that I was bothered by the fact that economists only recognized things that involved the passage of money, and that there were other things out there that were economic that weren't visible. And what I found was that the value of women's unpaid wages at that time was greater than all the wages paid by all the manufacturing companies in America. Of course that was the '70s, and it really wasn't very popular to talk about the value of women's work."

Scott Burns
Dallas Morning News

Scott has observed that time is more valuable than money and can never be equated with money, due to our finite human life spans. Many people today opt for less money so as to have more time, as Betsy Taylor's surveys show (chapter 1). Scott Burns believes that as more women get college degrees than men, it's not unimaginable that we could have a lot of Mr. Moms ten years in the future. "I know we're talking about the Love Economy, but let's talk about it more broadly. Where do we nurture people? What we're doing with families, with households is hollowing them out. We've heard of the hollowed out manufacturing company where the manufacturing company still exists but the manufacturing is done somewhere else. To the extent we monetize more and more, and devalue the nurturing that occurs in families, we're hollowing out that whole vital tool for moving the culture to the next generation."

Scott's recent work is focused on today's huge demographic shifts in the United States, Europe, and Japan as populations age. Much alarm has been raised in the United States about whether social security, pensions, and retirement systems will adequately support huge graying populations. Scott's view is more hopeful, "We keep talking about the retirement of seventy-seven million baby boomers as being a gigantic disaster and when you look at the money economics of it, it is. Yet there's an enormous 'glass half full' there. Seventy-seven million people are going to retire—what if they do something else? What if they say I'm going to reinvent myself, I'm not going to retire. What if they form the American Association of Re-invented

People—AARP? Seventy-seven million people can do volunteer work in schools, nursing homes, and other places. There's a whole potential economy out there." Luckily, some mainstream economists have ceased much of their doom mongering as healthy, vigorous retirees, find whole new careers and new meaning in all kinds of volunteer community service. The current pension mess in the United States involves an unconscionable shift by companies of their pension liabilities to the taxpayers, a move that has sunk the federal Pension Benefit Guaranty Corporation under $87 billion of such liabilities over the next ten years. Contrary to the dire predictions of many economists, people are living longer, healthier lives and are choosing to work and have second careers in volunteering long after usual retirement age. *Business Week*'s cover story "Love Those Boomers" (Oct. 24, 2005) sees them as a vital seventy-seven million—27.5 percent of the U.S. population with $2.1 trillion of annual spending power, many of whom are finding new meaning and service opportunities in their communities.

Rebecca Adamson, founder of the First Nations Development Institute, Fredericksburg, Virginia, understands both the Love Economy and the money economy and has found many creative ways to reconnect them for healthier community development. Rebecca points out, "Within all traditional societies there is actually a sense of enoughness . . . from the most remote Alaska village to the most remote Sami camp fire. I have been in places where we're basically going to eat what we hunt. Beavers mate for life, and one time we were in a village and one of the fellows shot a beaver. We cooked it and we were eating it when the mate came out looking for him. At that moment you

Rebecca Adamson
First Nations Development Institute

had more respect for the life that you took so that you could survive than any kind of giving of thanks. But while the mate is looking for its partner, one of the kids picked up a gun and was going to go get it. Every single elder in the village said, no, we have enough! And it is an absolutely operating principle that I've called enoughness. I think that there's wisdom within all these societies that sustained themselves over generations and generations for twenty, thirty, forty thousand years with a sustainable principle that I would absolutely say is enoughness." Rebecca helped design The Lakota Fund, which links investors and donors with Native American communities in mutually beneficial networks and exchange. She says, "We can do this. I know community economics can once again use these ancient, ancient principles and have capital connect to community."

Social scientist Riane Eisler, author of *The Chalice and the Blade* (1987) and *The Power of Partnership* (2002), also focuses on the unpaid, but indispensable caring sectors of our economies. Riane is cofounder of the Center for Partnership Studies, Carmel, California, which researches quality of life in many countries. Riane notes, "Most people including policy makers when they think of quality of life, if they even think of quality of life rather than just the single measure of GDP, they don't think of women. It's just outside of their landscape, outside of their frame—but it needs to be inside the frame. We did a study at the Center for Partnership Studies in which we took two clusters of measures. One—general quality of life. What does that include? It includes everything from infant mortality to availability of potable water to human rights to how large the gap is between those on the top and on the bottom of income. And then we took the other cluster, which was measures

Riane Eisler
Center for Partnership Studies

of the status of women. And what we found is that in significant respects the status of women is a better predictor of general quality of life for men, women, and children than even GDP." Riane then drills deeper into her findings, "For example, Kuwait and France at the time had almost identical GDP. But that most basic measure of quality of life, infant mortality, was twice as high—double—in Kuwait than in France. And of course the status of women is much higher in France."

Riane cites another example, "In the United States, those professions that do not involve care giving, plumbing, engineering, are much higher paid than those that do. So you find this ridiculous thing that people in the United States will think nothing of paying $60 an hour to the person to whom we entrust our pipes but for a childcare worker $10–15 at the most. That work must be included in measures of economic productivity. The good news—there's movement in that direction. The Nordic nations and Canadians are beginning to do it. One of the big issues, of course, is how do you value it?" Riane believes that economics is basically about what we value, and we must value the work of caring and care giving and develop caring policies—the subject of her book, *The Real Wealth of Nations* (2006).

Edgar Cahn
Founder, Time Dollar Institute

Another inspiring pioneer of the Love Economy is lawyer, Edgar Cahn, creator of Time Dollars, or more generically, Time Banking, explained in his *No More Throwaway People* (2004) and his *"How To" Manual: The Time Dollar* (2004), used worldwide by community groups to set up "time banks" run with personal computers to keep track of exchanges of members' services to each other. Edgar jokingly explains these local exchange systems and their huge, unnoticed value to national economies. "This kind of community economy doesn't produce much of significance; it just produces children, healthy families, safe neighborhoods, care for the elderly, democracies, civil rights, and social justice movements. Economists who have looked at it concede that it's at least as large as what's measured in the GDP and in some countries it seems to be even larger. When they did a study of the unpaid care that keeps seniors out of nursing homes it worked out at $8 an hour to $196 billion each and every year. That's six times what we spend in the market economy on services for seniors, that's three times what the federal government spends on nursing homes. We're not talking about a minor economy, we're talking about a major economic system that's off the map."

Recently, economists have begun to call core activities such as these social capital. We prefer the Love Economy. While economists still count homemakers and stay-at-home moms and dads as "unemployed and not economically active," the 1995 United Nations *Human Development Report*, mentioned earlier, found that this and all other unpaid work amounts to $16 trillion missing from the official global GDP figure of $24 trillion. Represented women's unpaid work was $11 trillion and $5 trillion for that of men. Even today, global economic statistics still exclude two-thirds of the world's output. Unpaid work has been studied by many sociologists using time budgets, which estimate hours each day spent in growing food, building housing, cooking food, educating children, caring for the young, elderly, and disabled, volunteering in communities, digging wells, building schools and clinics—time-honored activities that do not involve money. Billions of the world's people still engage in such activities, including those women working in families and households everywhere.

Edgar Cahn helps us see how time can be valued. "You put in an hour, you get an hour's credit then you can spend it to get help for yourself or your family or you can give that time credit to someone else who needs it more. We have a time dollar broker whom you can call—so you're not asking for charity. If you don't pay back, your kids can pay back or somebody in your family can pay back. That model we've come to call a Neighbor-to-Neighbor model. People don't like to share their problems but they don't mind calling a broker and saying, Do you have somebody who can take care of my dog this weekend? Nobody's going to go up and down the street saying, I'm going away this weekend, could you take care of my dog? So, with a broker to make those arrangements we can begin to rebuild trust amongst neighbors who don't know each other or who know everything bad about each other but nothing good about each other. So we're rebuilding a sense of extended family in those communities."

Edgar offers another example of how time can be valued and rewarded. "In the Chicago public schools we were given five of the worst schools in Englewood which was known as 'a killing zone.' We were asked, 'Could you get kids learning?' I said, 'Yes we can.' Give me some fifth or sixth graders who will put in one hundred hours and who can earn a recycled computer. After a year in the program the tutor and the tutee averaged a year's gain in both math and reading. At this point about 4,500 kids have earned computers. We're redefining as work and honoring the work that it takes to raise children, to build strong families."

Time-banking examples can be found in Japan, Korea, Spain, and other European Union countries (www.timedollar.org). Worldwide, we see the value of human time, the societal value, satisfaction, and happiness it affords. As discussed in chapter 1, sociologists study happiness and they find that it is not directly proportional to income but it can be a result of self-sufficiency and other factors. For example, small-scale farmers are often outside the money system, so their contributions to the economy are frequently overlooked.

Vandana Shiva, founder of Navdanya in India, is an Indian physicist who has inspired millions by her commitment to defend the rights of women in rural Indian villages and small farmers to protect their local watersheds, forests, and native species. She has taken up the cause of endangered farmers worldwide. Dr. Shiva says, "I think one of the biggest illusions of our times is that human welfare, human well-being comes out of the money that is a pure fiction, a piece of paper that says I promise to give the bearer goods worth whatever." Vandana has authored many books, including *Staying Alive* (1989). She describes her organization's main work saying, "We are working to save biodiversity, native seeds with farmers communities, and create community seed banks. We also work with them to promote organic farming. I have seen well-fed communities with no hunger, no thirst, no poverty with no cash economy, and they are very, very poor people. So you can have hunger with lots of cash flow and zero hunger with no cash flow. I think it's time for us to turn to those other economies of nature and sustenance and in the sustenance economy it's primarily women who play a very, very big role. Those economies are what keep life going on this planet, those are the ones that get rid of hunger, that get rid of thirst, that get rid of alienation, that get rid of insecurity. I have never seen terrorism and extremism emerge in societies where people were left their freedoms to generate their livelihoods and provide for themselves." Many people support these local farming communities as part of maintaining cultures and biodiversity. In industrial countries people can also support these efforts by saving their own seeds and promoting farmer's markets in their communities.

Vandana Shiva
Founder, Navdanya, India

Susan Witt is the executive director of the E. F. Schumacher Society in Great Barrington, Massachusetts, founded in honor of the British economist and philosopher E. F. Schumacher, author of the best-seller *Small Is Beautiful* (1973). An expert on the Love Economy, Susan is the creator of the SHARE micro-credit program. It allows people who have the vision to create quality goods and services for local consumption, but who aren't able to secure traditional bank financing, the opportunity to secure funding for their ventures and grow their local communities. The society hosted the successful conference "Local Currencies in the 21st Century" in 2004 (www.localcurrency.org). Susan explains, "What the Schumacher Society does is advocate for a system where the goods consumed in a region are produced in a region. SHARE stands for Self-Help Association For A Regional Economy. It's simply a loan collateralization program. Citizens interested in participating go down to a local bank, open a savings account, which is a joint account with SHARE. Then those savings accounts are used to collateralize loans that the SHARE board approves but would normally not be approved by a bank. We call it an extension of the grandmother principle. We've just extended the number of grandmothers to the whole community." Susan cites a vivid example, "In 1989 Frank Tortoriello, who is the owner of The Deli, a popular restaurant in Great Barrington, lost his lease and had to move to a new location. He needed $5,000 to renovate the new location. We said, Frank, you don't need our extended pool of grandmothers because you have your own in your customers, borrow from them . . . and so was the birth of Deli Dollars." Freecycle.com takes another approach, offering thousands of usable, secondhand items free; the brainchild of Deron Beal, a charity worker in Tucson, Arizona. The site has 1.3 million users in over fifty countries.

Susan adds "One of the big misconceptions is that anything other than a federal dollar is illegal. Not true. The scrip need only be exchangeable with U.S. dollars so that the transactions can be recorded and therefore taxed. Deli Dollars inspired the creation of Ithaca Hours in upstate New York, (www.ithacahealth.org) by Paul Glover, who went on to create a healthcare system based on this local currency, and now sells kits to other communities on how to start and run a local currency."

Back in 1991 when Paul founded Ithaca Hours, he approached a group of other people in the local grocery store and said, "I'm starting this new barter program . . . it's currency-based rather than time-based (like an earlier barter system in Ithaca that failed). Each person received $20 worth of the money for free for joining and after that the *Ithaca Hours* newspaper shopping directory of offerings grew slowly over the next year. Soon businesses began accepting it and soon Ithaca

Hours became an institution that works because it keeps money in the community a little longer than other money. Typically you spend a U.S. dollar and it leaves town after several transactions. Ithaca Hours helps build community because it gets people talking to each other and in some cases enables members to get interest free loans, as well as mortgages and healthcare. Prosper.com, a site similar to Friendster, where members can lend and borrow amongst themselves, is an online version of the age-old credit circles like Accion and Womens World Banking (chapters 5 and 7). Prosper's members have made fifteen hundred loans to each other totaling $7 million and spawned many micro-businesses (*Business Week*, July 3, 2006).

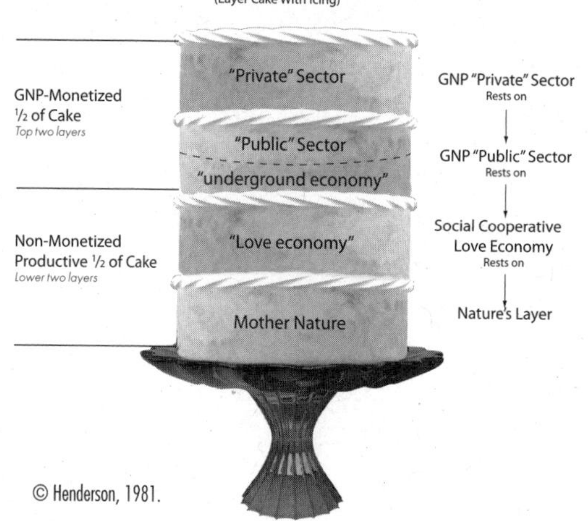

Today, barter—still disparaged by economists as primitive—has gone high-tech. eBay, the world's largest garage sale, demonstrates how mainstream markets can be bypassed. Many developing country governments barter oil, trucks, machinery, and etcetera, with each other bypassing the need for conventional foreign exchange. Such bartering is called "counter trade" and has been estimated at between 15 and 25 percent of all world trade, as documented in *Counter Trade, Barter, Offsets* by Pompiliu Verzariu (1985).

We've only glimpsed the world's unpaid economies still invisible to economists and their money-based statistics. The total production of societies looks like this three-layer cake with icing—broader and deeper than the economists' monetary pie. The icing on the cake is the private sector, with entrepreneurs creating new businesses and market-based companies and industries. The icing rests on the layer of the tax-supported public sector: roads, schools, sewer-systems, infrastructure networks from air-traffic controllers, and our military and government agencies that monitor our food supply, water, and air quality. These two top layers of our productive "cake" are conducted in money—generate our paid jobs and are tracked by economists in our gross domestic product (GDP) and other money-based indicators. But the two bottom layers are mostly overlooked—missing from GDP and corporate balance sheets. The loving, caring, sharing work within households and communities—those thousand points of light—is the missing half of a country's production and exchange. The bottom layer is nature's productivity, which underpins all human economies—our basic life-support system. When public officials and private decision makers are blind to these two lower layers that underpin our GDP, they miss many creative options for fuller employment, enhancing safety nets, reducing crime and drugs, revitalizing small towns and inner cities—at less cost to taxpayers!

Let's clarify the difference between money and wealth, GDP-growth and quality of life. All are important! Let's connect all these dots! As we saw, Time Dollars, Deli Dollars, Ithaca Hours, and *Barter News* all show the truth: information and money are equivalent. In *Building a Win-Win World* (1996), I showed that information is the world's new currency and it isn't scarce. Even Citibank's former chairman noted in the 1990s that the world is no longer on the gold standard—but on the information standard. Pure information-based trading, such as on local radio stations where farmers exchange seeds, tractor time, and other goods and services, is the wave of the future as I discussed in my "Addressing Key Causes of Poverty and Inequality" speech in March of 2001, at the Inter-American Development Bank Annual Meeting in Santiago, Chile. Today, better information, all the new indicators, help us see our economies whole—as living, creative expressions of all our values, and the vast real wealth and potential of all our people.

ROUNDTABLE
WALKING THE TALK

Simran with Bob Meyer

Hewson Baltzell

Simran points out that in the Love Economy, we're looking at a movement instead of focusing on a single company. We examine Barter News, a firm founded by Bob Meyer, which tracks barter activity and demonstrates you don't need money to make the world go 'round. Simran and stakeholder analyst Hewson Baltzell interview Bob Meyer to get a better understanding of how barter impacts our local and national economy.

Bob Meyer: The scope of barter in the United States is probably larger than you expect at about $15 billion a year.
Hewson: Why does barter exist at all in a modern money based economy? What do you need barter for?
Bob: Barter exists because it's a conduit to another area of currency, a barter currency. You've got your cash marketplace, you've got your

barter marketplace, and they don't compete with one another. It's an adjunct, an addition. Additionally, when you trade you do so at your variable cost of doing business because your cash business takes care of all your fixed costs. So it's a very profitable way of doing business. Let's say you have some extra products, which you haven't sold, they're sitting there in inventory. You can do it on a direct basis if you find somebody who wants your product and you want theirs—that's a little tougher to do—or you can do it on an indirect basis. This is growing here in the United States as well as worldwide. There are six hundred companies in the United States now that are trade exchanges. A trade exchange has their business card but it's only good within this closed network. This firm is in Orange County, where I'm from, in California. There are fifteen hundred businesses that are members of this particular company. So, when I sell something and I earn trade dollars it goes into my trade account. And then when I want to go down to the restaurant that's a member of the trade exchange, I pull my card out and I pay for it with my trade dollars.

Simran: I'm sure the Internal Revenue Service wants to make sure they're getting a share of all this. What are the tax implications of bartering?

Bob: Well, in 1982 under the TEFR Act—Tax Equity Fiscal Responsibility Act—barter companies, trade exchanges, if you will, were classified as third party record keepers and they send a 1099 to the IRS the same as credit card companies, and the same as banks—any third party record keeper.

Hewson: What are the biggest items that get traded? Is it primarily retail?

Bob: Everything through a trade exchange is done at a retail basis. So retail prices, same as you sell for cash you sell for trade. And everything under the sun, like I had orthodontic work done for my two boys—$10,000; The dentist wants the trade dollars so that he can advertise in the local community papers to get his name out to bring in more cash business.

Simran: Are there any kinds of codes of ethics that regulate these kinds of transactions and are there any socially responsible practices or other initiatives like that?

Bob: This is a closed system that really walks the talk. We have two trade associations that regulate the trade exchanges, if you will—the six hundred barter companies in the United States. They have a code of ethics and you have to do certain things and rely on certain rules, and what not, and if you don't you're expelled from the organization.

Hewson: Can a small community-based business benefit from barter?

Bob: Yes, through trade exchanges. About 2 percent of the businesses in the local community will be members of a trade exchange and on average they will do about $6,000–$10,000 a year in business.

Simran: What about people who don't own businesses?

Bob: Well, let's say you bake delicious pies. There are local organizations in many cities around the country now that are what I call socially responsible bartering organizations. They're not set up like a trade exchange, per se, in that they have businesses that are doing quite a significant amount of barter, but they do offer the person an opportunity to trade their pies and maybe have them come over and do their gardening.

Hewson: How much of your own business is done through barter?

Bob: I do about 60 percent of my business on a noncash basis. Most of the work that I do is outsourcing the production of my publication. And what I do is trade on a part cash and a part trade basis. I know businesses that are doing a million dollars a year in barter through these trade exchanges. Now that's high but the point is that they wouldn't be doing this much, if it wasn't profitable and advantageous.

FOUR

Green Building and Design

IMAGINE THE MOST HIGH-TECH BUILDING POSSIBLE, and then consider this possibility—designing a building that generates oxygen, distills water, provides a habitat for thousands of species, collects solar energy as fuel, grows food, and looks beautiful. Sound impossible? Well, it's not. "Five years ago, you would mention green building and get a lot of blank stares," says Alex Wilson, editor of the monthly newsletter, *Environmental Building News*. "Today it's a known term for an increasingly large portion of the population." Alex is president of Building Green, Inc. and he offers some easy steps anyone can take to green their own house inexpensively—and these changes can save money, too. (www.buildinggreen.com). The University of South Carolina recently opened the world's largest green dorm in Columbia, South Carolina, built with much recycled material, from cement blocks and copper roof to interior carpeting. Students and faculty help grow and plant drought-resistant, low-growing greenery as part of its water management system. The new West Quad dorm cost $30.9 million, no more than if built conventionally, and includes a turf roof, a learning center for its five hundred residents that is partially powered by a hydrogen fuel cell, and a café that sells healthy foods and environmentally sound products. West Quad uses 45 percent less energy and 20 percent less water (heated by solar energy) than other dorms and is free of ozone-depleting substances. With much natural light and high-efficiency washers and dryers, the 172,000 square-foot complex shows that students can live comfortably in an environmentally friendly facility that also saves resources and money. So why haven't buildings always been built this way? Because the same faulty economics that blinded companies to saving money through efficiency and smarter production and design also fooled architects and builders.

Adam Joseph Lewis Center
for Environmental Studies at Oberlin College

Sustainable architect William McDonough has drawn inspiration from nature in designs for clients like Nike, Ford Motor Company, Gap, and the Smithsonian Institute. Bill is the founder of William McDonough and Partners, Charlottesville, Virginia, and is a preeminent green architect worldwide. He explains his firm's approach: "Our goal is a delightfully diverse, safe, healthy, and just world, with clean air, water, soil, and power, economically, equitably, ecologically, and elegantly enjoyed—period. It's a fundamental question of quality. How can a building be a delightfully elegant place if it makes you sick, or destroys the planet? What we are trying to do is create human ar-

William McDonough
Architect

tifice in the same framework of a living thing. So if you look at what it means to be alive, you realize you have to have growth, you have to have free form of energy in order to expand, which comes from the sun. And you have to have an open system of chemicals that move through metabolism. So, if we look at that as the fundamental mechanism and design framework for biology, we're saying how would we take that and apply it to technology. Natural systems would only support about four hundred million humans. So we need synthetic activity so we can support the other 5.8 billion. So we need materials in closed cycles. So what that means to a building is that we ask the question, how can we make a building like a tree? And how can we make a city like a forest, so that it's fecund?" Bill's ideas are leading a green design revolution as pictured in *Business Week*, June 12, 2006, which also features a profile of McDonough.

Ford's Green Plant

Hunter Lovins, who wrote the foreword to this book and coauthored the landmark text, *Natural Capitalism* with Amory Lovins and Paul Hawken, is another pioneer of green design. Hunter heads Natural Capitalism Solutions in Colorado and teaches courses on sustainability and eco-efficiency. She was named a "Hero of the Planet" by *Time* magazine. Hunter observes, "People like Bill McDonough say that eco-efficiency is actually wrong. What we need to be is eco-effective. Just using less,

Hunter Lovins
Natural Capitalism Solutions

being less bad, doesn't equal being good. And he has a point there. It enables a company to save money, to experiment with doing things in a slightly different way, but it's only a first step, it is the first principle of natural capitalism. The second step of natural capitalism is green design, but even then we have to manage all our systems in ways that are restorative. That's the third principle of natural capitalism. I think if you do all of these over a period of time, we will get to a point that will pretty closely approximate the best set of criteria for a system to be called sustainable.

Sustainability—now an often-misused buzzword—came into use in 1989 with the release of the U.N. report on sustainable development, *Our Common Future*, chaired by the prime minister of Norway, Dr. Gro Harlem Brundtland, MD, who later became head of the World Health Organization. The report defined sustainable development as "development which meets the needs of the present without compromising the ability of future generations to meet their own needs." Bill McDonough took this definition to heart, applied it to all of his work in green design and architecture, and spelled it out in his book *Cradle to Cradle* (2004). He describes his work for the retailer, The Gap, at their corporate campus, in California. "The roof is an undulating meadow of the ancient grasses, so we had to get permission to get seeds from native lands to put them there. It's a restoration act. We were trying to design a life support system for people at work. Not a work support system for people who don't have a life. The building is full of fresh air. We use the nighttime to cool that building down and it's full of daylight. So it's really an optimized work place and that is why the clients have taken it up with such delight. They get massive productivity increases and if I can get a 4 percent productivity increase, which is only ten to twenty minutes a day, I can pay for the entire building. If I got two minutes a day, I pay for all the premiums for all of these marvelous effects." Today, many companies like the Gap, Ford, Toyota, Nike, and others see new green corporate buildings as image-enhancers. In many cases, such corporate green makeovers are a start in the right direction but don't substitute for the needed top-to-bottom redesign of products, manufacturing, and all other corporate operations worldwide.

Kathleen Hogan directs the highly successful Energy Star program at the U.S. Environmental Protection Agency. This label is familiar to American consumers looking for the most energy efficient—and therefore cost-saving—appliances on the market. Kathleen explains that "The Energy Star program is a program of the Environmental Protection Agency to give people the tools and information they need to make sound energy decisions. What that really means is to choose energy efficient solutions. The Energy Star label is on about forty different types of products at this point in time. Many of them go into the homes—the heating and cooling equipment, the appliances in your kitchen, home office, home entertainment. The Energy Star label, when it's on a product, means that it will save you money, it will save energy, it will help protect the environment, and it provides you all of that while maintaining or enhancing the performance of the product, so it's just a good all around symbol."

Kathleen Hogan
U.S. Environmental Protection Agency

A typical Energy Star home uses 30 percent less energy than a conventionally built home. Habitat for Humanity, spearheaded by Linda Fuller, has a New York City affiliate who ensures that the people who need cost savings the most are getting the benefits of Energy Star-rated appliances. A fifteen hundred square foot, typical single-family Habitat home has a parlor floor with the living room and the kitchen, which brings together many of the elements of an

energy efficient and healthy home. Habitat for Humanity has had a green building program for about ten years and uses the basic Energy Star standard for all its homes across the United States. In New York City, they have been building Energy Star housing since 2000, with a conscious effort from the design process all the way through to the management of the building. This includes increased insulation, very efficient boilers, healthy and nontoxic paint finishes in the building, and that the building works as a system to reduce energy cost. Maria Sanders, a happy new homeowner, enthuses, "I was watching Bob Vila on TV and this particular show he was building with Habitat for Humanity. I looked at the faces, the enthusiasm these families had in building their own homes. I turned to my son and said, We could do that! I could do that! We could build a home!" And they did!

NYC Habitat Homes

Kevin Sullivan, formerly with Habitat for Humanity in New York, explains their philosophy, "A home is about much more than bricks and mortar. It's really about people's livelihood and lifting themselves out of poverty through their homes. Through green building, we find that through a healthy home that also puts money back in the pockets of families, they can lift themselves out of poverty sooner. They can make sure that their kids are healthy, and that they have a real, long-term investment in their future. Green building is about the bottom line, so it's absolutely good for everybody, whether you're saving 20 percent of your income through utilities or whether you're saving 2 percent. More importantly I think it creates the notion that we are all connected to each other through our communities and how we use resources." Much of the development within green building has been driven, in fact, by educated consumers demanding Energy Star appliances and homes. In fact, the National

Association of Home Builders' 'Green Guidelines' are being pushed by customers. It means higher resale value on homes, lower maintenance costs on homes, and it's a real benefit for any family that wants to undertake this. But we have to be very conscious of the environment in which we work. If we build McMansions and we throw in a little bit of Energy Star appliances, that's not enough. Hunter Lovins adds, "If you build a building you can ask for LEED (Leadership in Energy and Environmental Design) certification if your building is energy efficient, if your building is a healthy building to be in, uses relatively few resources. Look up U.S. Green Building Council, and get the LEED standards at www.usgbc.org/LEED/."

Yet, those economic textbooks still need an overhaul when it comes to measuring energy efficiency because so many social costs and impacts of energy use were externalized from producer balance sheets. Kathleen Hogan sums up why energy efficiency is vital to strengthen the economy. "Energy is one of these things that's not tracked all that well within business. And that really makes you ask the question: 'How can you manage what isn't very well measured?' So we've designed a brand new, innovative building performance rating system. So you can rate a building on a scale between zero and one hundred. If it's close to zero, it's an energy hog, if it's close to one hundred it's an exemplary building for energy efficiency. If you're down toward that lower end, there's just a tremendous opportunity for cost effective improvements in those buildings. People that have purchased Energy Star products or improved their buildings are saving about $8 billion worth on energy bills. They're avoiding greenhouse gas emissions equivalent to about eighteen million cars, and we're saving the amount of power required for about twenty million homes."

Meanwhile, the United States is still struggling to become less dependent on foreign oil and reduce use of fossil fuels, whose combustion creates many pollutants as well as the rising levels of CO_2 now contributing to climate change and global warming. Now, there are widespread calls for a Manhattan-type project to shift toward renewable energy—solar, wind, harnessing ocean currents and biomass, and further develop innovative technologies, including fuel cells and flexible fuel vehicles, such as the Apollo Project promoted by labor unions and environmentalists. Powerful lobbies of the fossil-fueled industrial sectors still hamper these efforts, even though, as I described in *The Politics of the Solar Age* (1981, 1988), this huge transition is inevitable for all societies. Change disrupts existing industries and special interests, which explains why it has taken so long. Moreover, multibillion dollar subsidies have been conferred on coal, oil, gas, and nuclear power companies. The technologies of the emerging solar age have been waiting in the wings for decades, some for centuries. These technologies are all about information and light. Finally, momentum is building to launch and finance these technologies through support for thousands of start-up companies represented at such meetings as the Cleantech Venture Forums (www.cleantech.com) founded by Nick Parker. Green buildings are now all the rage, with an eighteen-page advertising section sponsered by major companies in the June 19, 2006 issue of *Business Week*.

Now we have green skyscrapers, including the Swiss Re tower in London, Pittsburgh's new convention center, the Conde Nast building, the Bank of America Tower, and other buildings in New York. The Earth Pledge Foundation in New York demonstrated the feasibility of green building by converting a Manhattan townhouse. Its Green Roofs Initiative has encouraged many others to make this a global transition. Leslie Hoffman, executive director of Earth Pledge, describes this wonderful house. "We've overcome the false idea that environmentally friendly means less-effective, less aesthetically pleasing, and more expensive. So we've showcased the best of contemporary design and the best of environmental technology." Leslie has a degree in architecture and

Leslie Hoffman
on Earth Pledge's Green Roof, NYC

design and was a custom residential green builder for many years in Maine starting in 1979 until 1990. She's been the executive director of Earth Pledge since January of 1994. "In the late 1990s, we acquired a townhouse originally built as a residence for Abraham Lincoln's granddaughter. Earth Pledge decided to do an eco-renovation on the building. Small control panels throughout the building control all of its heating, lighting, and air conditioning for energy efficiency—with wired in with motion sensors. All fans and lights go on and off automatically."

Earth Pledge's Green Roof, NYC

Leslie elaborates on how her philosophy meshes with technology, "We use this building as a way to inspire people to think about the choices they make as they build new buildings and as they retrofit old buildings. We have a green roof, the first highly engineered, lightweight green roof system on a building in New York City. These green roofs mitigate several very important environmental problems including urban heat island effect, combined sewage overflow, while at the same time bringing greenspace. All these are valuable for biodiversity and human pleasure, but also save energy and extend the life of the underlying roof membrane by as much as two to three times. All of the materials and techniques that we used were thought about in terms of environmental footprint, whether it was from which companies we were buying products, to how energy efficient those products are, to the indoor air quality of the building, to the recycled content of some of the materials—fabrics, fibers, furniture parts, etcetera."

Chicago and many other cities now boast green buildings. Bill McDonough is now designing an entire green city from scratch in

China. New companies spawned by green philosophy include New York's Consumer Powerline, which enables apartment buildings and offices to make money by saving electricity. They recently signed up Macy's and the Starwood Hotel chain worldwide.

Green design grew out of the scientific insights of ecologists, biologists, and systems thinkers who gradually came to realize that economics had developed without these crucial interdisciplinary insights into the functioning of the natural world—our ecological life-support system. Ever since Rachel Carson's *Silent Spring* (1962), economists and businesses have fought a rearguard action—hoping not to add the social and environmental costs of their products to company balance sheets and their products' prices. Today, there is wider acknowledgment that full-cost prices (which reflect all previously externalized costs) are the key to steering consumer societies toward sustainability. London-based Truecost helps companies calculate these full-cost prices, redesign products, and identify opportunities to save resources and money. The VIA3 network (www.via3.net) also in London, which I cofounded with John Theaker, CEO, operates greenyouroffice.co.uk, which electronically matches green office suppliers with companies seeking their services.

Dr. John Todd is an internationally recognized biologist and a visionary leader in the field of ecological design. John and his wife, Nancy Jack Todd, founded the New Alchemy Institute, which pioneered "living machines" that treat waste to produce food and fuels, and restore damaged aquatic environments all by following nature's example. John Todd is now a professor at the Rubenstein School of Environment and Natural Resources at the University of Vermont. He describes his basic insights: "It's kind of decoding the language of nature and then creating from that language, if you will, blueprints for design. It's called ecological design or, as we originally called it, 'new alchemy.' The new knowledge would have to be integral to all the disparate parts of human enterprise, embracing systemically food growing, energy capture and use, building and building design, landscape management, even transportation. Somehow all had to be linked together in a new whole. At Ocean Arks, we began to find out if it's possible to create living technologies that would start to heal the damaged waters, clean up their pollution, and provide protection for drinking water reservoirs." John uses a hypothetical example. "Let's say we want to make pure water out of sewage. The next thing one does is assemble all these life forms. You put those in giant vats, connected together like beads on a string, and in each vat you put huge numbers of species, hundreds, maybe even thousands of species of life forms. Then you introduce at the top end, the waste, and it starts to flow from one vat to another, down step by step, and so in this case there would be ten stages of transformation of the waste to pure water. As the ecological engineer you need to provide them with oxygen support, in most cases, and after that they get down to work. You also have to give them the intelligence of the seasons—And the way you give a living technology the intelligence of the seasons? This is obvious when you hear it, but it's not very obvious *until* you hear it. You go out into nature at every season and gather the life that's active at that season and put it in. So, all of a sudden this system, let's say it's inside a greenhouse, has the wisdom of winter, and the wisdom of spring, summer and fall."

We humans learn from Mother Nature, the planet's preeminent designer. Human technologies mimic nature. Industrial designers openly search natural forms, from abalone shells to spider webs, to learn how nature performs her design miracles.

This kind of deep systemic awareness characterizes the mindsets of the best designers and builders. Its humility is its genius—remembering that we humans can learn from Mother Nature, the planet's preeminent designer. All human technologies mimic nature in some way. The best industrial designers now openly search natural forms, from abalone shells to spider webs, to learn how nature performs her design miracles. Such new insights are described by Janine Benyus in *Biomimicry* (1997). Kevin Sullivan of Habitat for Humanity has also learned these deeper lessons, "I think the larger point about green building is that we're all tied together, we're all in this lifeboat together and we're consuming a non-renewable resource more and more. We can really reverse the long-term trend towards being net consumers of energy. Houses can really be thought of as producers of energy and we can really change the way we live and interact with the environment. Then long term, we can reduce our impact on the planet, and feel more connected to each other and also to our sources of energy and materials in the world."

Bill McDonough offers similar sentiments, "We need new consciousness. It will be how to combine nurture with opportunism. And so what we are seeing on the part of the global companies is that finally a revelation that local communities are full of diversity, and biology, and delight. And the human culture needs to express itself on a local level. All sustainability will be local. We want four hundred kinds of French cheese, not one kind of cheese! We want full diversity in biology. But globally, we want global standards and the global companies will maintain their license to practice as long as they keep raising the standard of living for everyone."

Hunter Lovins agrees. "Capitalism as presently conducted is, if you will, violating its own internal logic. We're not being very good capitalists, because we are losing the human and the natural capital, which are at least as important to economic well-being as the forms of capital that we do count. For example, a rough estimate of the value to the economy of natural capital comes to about $30 trillion a year—about the same as all the economy that we do count. And if we're losing this, there ought to be some indicator that says to a company, to a community, to an economic system, you're actually getting worse off."

Today, many hotel chains are following Fairmont's lead to greater sustainability, better health for employees, and more savings at the bottom line. We have only peeked at the growing numbers of cleaner, healthier environmentally friendly office buildings and houses now available and seen that they are also more beautiful and enjoyable places in which to live and work. These better methods of improving overall quality of life and increasing productivity are driven by urgent problems: rising costs and dependency on imported oil and polluting fossil fuels. Energy-conserving buildings can tap the vast reservoir of energy the United States wastes each year. Energy expert, Dr. Skip Laitner, shows that consuming our energy more efficiently could substitute for most of the oil we import—and cut pollution substantially (www.ethicalmarkets.com). Healthier buildings and less-polluted urban areas can lower health bills, too. Lower up-front costs of putting up inferior and "sick buildings" are shortsighted when long-term overall costs are calculated. More economists now account for such costs and benefits, calculating maintenance and other costs over the buildings' life. This leads to better siting of buildings to capture the sun's energy for heating and cooling and smart windows that allow more natural light. The higher standards in building construction, and the Energy Star label on refrigerators and other home appliances and office equipment are leading the way. Working with their local builder members, the National Association of Home Builders and the U.S. Green Building Council promote nationwide green building standards. Homes meeting these standards will be marketable as green-built. These criteria include:

Land development and siting

Resource-efficiency

Water conservation, an increasing imperative worldwide

Indoor air quality

Energy efficiency

Here's some steps you can take to save money, energy, and reduce pollution:

1. Buy compact fluorescent light bulbs
2. Install water-saving shower heads
3. Tune up your heating and cooling units
4. Look for the Energy Star label and the new green-built standards

Consumers, employees, homeowners, builders, and materials suppliers, together, we can all do our part for a greener, cleaner, healthier future!

ROUNDTABLE

WALKING THE TALK

Simran Sethi

George Terpilowski

Simran hosted George Terpilowski, General Manager of the Fairmont Hotel in Washington, D.C., and Alice Tepper-Marlin, our stakeholder analyst, to discuss the choices hotels make. They are particularly important and they have an enormous impact on purchasing, energy efficiency, water conservation, and waste minimization. Fairmont is a Canadian-based chain recently purchased by a United Arab Emirates company, and it's in the process of greening, through its Green Partnership with forty-four hotels located across North America—Canada, the United States, and then through the Caribbean and one in Dubai.

George Terpilowski: Many of those hotels are located in pristine parts of those particular countries from the Canadian Rockies to the oceans of the Atlantic. The hotels are historical landmarks in many respects, so not only are they places where people come and stay overnight but they're buildings that have relationships that are very strong with the communities in which they're founded. One of the objectives of our company is to make sure that not only are those hotels prosperous

within those communities, providing employment and enjoying a symbiotic relationship, but also the companies and the hotels themselves contribute to maintaining the quality of the environment and the beauty of the locations in which they're founded.

Alice: What does that mean for you here in Washington, managing a hotel?

George: Here in Washington, D.C., we happen to be in the middle of a big city and the company has a Green Partnership program that is in place in all of the locations. Each of the locations adapts the corporate program to its individual geographic location. For example, we purchase portions of our electricity from wind-generated services out of West Virginia, limiting the amount of hydrocarbons that are distributed back into the environment. We encourage our employees by providing them Metro checks, which allow them to use public transportation instead of piling into cars and adding to the pollution that's caused in and around the beltway.

Alice: What do you do to buy locally, to support organic agriculture?

George: We just recently renovated one of our restaurants and reopened it—it's called Juniper. We wanted to make it an authentically local destination. One of the ways we try to achieve that is by pulling the indigenous produce and farm products that can be found in the Mid-Atlantic states. And then as the food is produced and eaten and then obviously on some occasions it's leftover, we then take the leftover product and we try to put those back into the community, most notably to food banks and soup kitchens that can be found in the city.

Alice: What about the way the hotels are kept clean? A lot of cleaning products that we use at our homes—brands that a hotel uses in very large quantities—endanger our water supplies and have toxic materials in them that are harmful for the environment and also for the cleaning staff.

George: One of the most elaborate programs we have is that of the Fairmont Orchid on the Big Island of Hawaii. The housekeeping department there has instituted a program where they've removed all of the chemical products that are traditionally used in the cleaning of hotel rooms and replaced them with organic products that are based on natural oils, lemongrass, thyme, vinegar, things like that. And one of the interesting by-products of this is that the housekeepers who work very hard at cleaning rooms found that they have much fewer allergic reactions, headaches, and things like that, that are probably attributable to the unhealthy and corrosive nature of some of the chemicals that are traditionally used in the hotel business.

Alice: What does Fairmont then do to take that best practice in Hawaii and bring it to the attention of the rest of the hotels?

George: All of the hotels submit the efforts that they make in terms of their own individual environmental programs to our corporate offices in Toronto, to the people that manage our environmental programs. Once a year, the company symbolically distributes trees to each of the hotels based on the restoration of natural products back into the environment that each of the hotels has undertaken. Part of the annual program is that the best pilot programs are put into a Best Practices binder. Then this binder is sent back to each of the green teams in each of the hotels. Many of the ideas that we've had, we've actually gotten from other Fairmonts.

FIVE

Community Investing

Globalization of finance and technology has impacted local economies by depleting environmental resources through unsustainable business practices. Scholarly reports, including *New York Times* writer Louis Uchitelle's *The Disposable American* (2006), talk about the growing wave of layoffs of workers in the United States and the disruption of lives and communities through outsourcing of production to low-wage countries. Global corporate giants often force out local businesses, including small farms. Still, in spite of all these new forces, your best investment might be right outside your door.

Current economic policies don't measure the deeper value of cohesive communities, families, and local cultures and thus overlook how they underpin the wealth and well-being of all nations. Twentieth century economics' false promises led to the massive human migrations in search of jobs and better lives, from rural areas to cities, where 50 percent of the world's population now live—often in terrible slum conditions. The promise of jobs fell far short and the huge costs of building city services and infrastructure led to massive government debts and huge unmet needs. Now a growing countertrend finds many people fleeing metropolises in favor of rural micropolises, smaller towns like Mount Airy, North Carolina, as described by Wanda Urbanska in her popular PBS TV series, and by Rich Karlgaara in *Life 2.0: How People Across America Are Transforming Their Lives by Finding the Where of Their Happiness* (2004). Local economies are still undervalued until they break down. At that point, the huge costs required for social services, unemployment, and drug and crisis counseling—and caring for homeless people—often fall on taxpayers. In this chapter, we meet a group of pioneers who recognize the wealth of opportunities for new jobs, thriving businesses, and residential developments that community investment affords.

Indications that a new trend toward local control and local community building abound. Ken Bohnsack's Sovereignty Bill, introduced into Congress in 1999 as HR1452, the State and Local Government Empowerment Act, takes another approach. The bill would restore to Congress the right to issue special interest free loans to local governments directly for public projects approved by voters, including schools, roads, bridges, public lands, mass-transit, and environmental enhancement—rather than the bonds localities have to float to private investors at high interest rates (www.interestfreeloans.com). HR1452 aims to put the power back into the hands of the local government in an effort to support local living economies.

Local living economies are created when individuals take action to restore and transform their community. Judy Wicks is one such leader. Her White Dog Café in Philadelphia is a robust example of a business that thinks global and acts local. The White Dog Café is a member of the Business Alliance for Local Living Economies (BALLE), which she cofounded to help community businesses drive their local economies. Philadelphia, like many older cities, has lost population, but its downtown is enjoying a rebirth, as described in *Edens Lost and Found* (2005), based on the PBS TV series.

Judy Wicks recounts how and why she started the White Dog Café. "I realized that the antidote to control by large corporations was local control. That's when I started to realize that what I needed to do was to expand on what I was doing here in Philadelphia. Buying from a local farm was the beginning. Having food security is the beginning, but there is also energy security. I started to envision a world where every community has food security, water security, and energy security. If every community had those things, that's the foundation for world peace! Then think what the other basics are: clothing, shelter, you know—the green building movement. I joined forces with a friend of mine, Laury Hammel, and the two of us cofounded BALLE. We work with twenty-five communities in North America including Canada where we have business networks that are doing the same kind of work that we're doing in Philadelphia. They also build local food security, support the local clothing industry, and support local alternative energy, and so on. Then we are finding ways that we can link between these communities. If you can't find something available in your own community, you look to buy it in a small business in another community, so that the purchase supports a local economy elsewhere. There are a lot of things that are not available locally. Coffee's not available locally, chocolate's not available locally. So this is a global economy, and this movement is not about buying everything local. It's about buying everything in a way that supports a local community where that project originates: in other words, paying fair trade prices. BALLE is not about money, it's about life. It's about supporting life. It's about fairness. It's about people. You don't make decisions just based on price. We've been taught we are suckers if we don't look for the lowest price, or that if we don't get the highest price if we are selling something, or we don't get the highest return on our investment, that we'd be suckers. In actuality, what we need to be looking at is not just the bottom line, but how does our purchase, whether we are a business person or a consumer, affect the world."

Judy Wicks and Friends
White Dog Café

William Drayton—a former administrator of the Environmental Protection Agency under President Jimmy Carter shares Judy's vision. Bill is one of the McArthur Foundation Genius Award winners, and with this grant, he founded ASHOKA (www.ashoka.org)—an organization that provides grants to roughly one thousand fellows—or social entrepreneurs—worldwide. These are innovative people whom Drayton believes will "make a scratch on history." Bill outlines his vision. "We need people everywhere to be change makers. Every primary social entrepreneur comes up with an idea that will change their field. For example, one of our Fellows in India created a telephone hotline that connects street kids to a telephone operator (also a street kid) to service them when there's a crisis. For the first time, supply and demand can get together. That changes how everyone does business. The policeman on the beat can no longer victimize street kids because accountability is one telephone call away, a free telephone call. This one entrepreneur has now made this work in fifty-eight cities across India; seven or eight million completed phone calls. It's now in forty-seven other countries."

William Drayton
ASHOKA

Josh Mailman understands the value of investing in community enterprises that increase both financial and social capital. He founded the Social Venture Network, based in San Francisco in 1987 and now a four-hundred-member group of venture capitalists, social investors, entrepreneurs, foundation, and nonprofit leaders, which helped spawn Business for Social Responsibility and many other organizations. Josh is also cofounder of Grameen Phone in Bangladesh with Iqbal Quadir (see chapter 7), a partner with the famous Grameen Bank. Josh says, "We have a huge concentration of wealth globally. There are individuals who wind up having huge concentrations of wealth. Maybe one day, through their own life experiences, they wake up and say, Hey, I really have to do something here, I am going to play a historic role in terms of making a major difference in the world with the fortune that I've inherited. That concentration of wealth is power that can be used to build global social movements, to support innovative models of change. I believe in the process of growth of individuals." The two founders of eBay, Pierre Omidyar, its chairman, and Jeffrey Skoll have both set up foundations to support local businesses and social entrepreneurs, the Omidyar-Tufts Microfinance Fund with Tufts University and the Skoll Foundation of Palo Alto, California, (www.skollfoundation.org). Bill Drayton adds, "Social entrepreneurs need a democratic society and they feed a democratic society. The two need one another. And the social entrepreneurs are very much the engine force behind the emergence and strengthening of democracy."

This global movement is linking socially minded investors and entrepreneurs in North America, Europe, Japan, Australia, and other industrial countries with their less-well-heeled counterparts in developing countries. This has led to the burgeoning of the global microfinance sector, which got its impetus from women, convened by the United Nations in the 1970s and 1980s, culminating in the Summit on Women and Development in Beijing, China, in 1996. Thousands of local credit circles comprised mostly of women entrepreneurs—including Women's World Banking and ACCION, which have been operating in many countries since the 1970s—came together to make sure that women would no longer be shut out of credit and finance. Professor Muhammad Yunus took the lead in Bangladesh and founded the Grameen Bank, which has lent more than $5.1 billion to over 5.3 million people. Grameen Shakti, a spin-off company, sells some fifteen hundred solar panel systems per month in rural Bangladesh and is growing 15 percent per year without subsidies.

Susan Davis, board chair of the Grameen Foundation in the United States, works with Alex Counts (chapter 2). Susan says, "If we can find more of these people, more social entrepreneurs who have innovative ideas to solve social problems, we can really accelerate the change process. What we do at Grameen Foundation is create partnerships with microfinance organizations and link them with individuals and organizations who want to work with them. Grameen Bank's average loan is now about $200; when it started it was about $67. They've made a profit every year but two when they got started and so this institution is a real model. It's a shining light for how people can lift themselves out of poverty. If you look at the largest program in India, three quarters of them have been able to bring themselves out of poverty after four loans or more."

Susan Davis
Grameen USA

Major investment funds are also launching community-investing efforts. The Calvert Group of socially responsible funds explains the need and benefit—both financially and socially—of these kinds of investments. Shari Berenbach is executive director of the Calvert Foundation based in Bethesda, Maryland. She describes their program. "The Calvert Social Investment Fund went to the shareholders and asked for permission to take 1 percent of the mutual fund assets and invest it, not into publicly traded stocks and bonds, but actually to take that money and make direct investments into non-profit organizations. These groups work at the community level in order to support affordable housing, to finance daycare, to support new jobs in small businesses. In 1995 Calvert launched the Calvert Foundation as a separate non-profit organization with a goal of popularizing community investment as a new asset class, as a whole new way to think about investing. We've done that by actually creating a special financial instrument called a Community Investment Note (like a Certificate of Deposit, or the CDs we're familiar with). When investors buy the Community Investment Note, every dollar invested goes to community organizations."

Rebecca Adamson, founder of the First Nations Development

Institute (chapter 3), is also a Trustee of the Calvert Group of mutual funds and explains further. "Community Notes—the idea really came out of the Lakota Fund, and it was the way that indigenous people would design a financial system, which first and foremost would be connecting capital to community." Rebecca helps native people from around the world protect their land and their rights. She adds, "In the case of both the Lakota Fund making small loans and Calvert, which was a mutual fund in the billion dollar asset range, the design principles were absolutely the same. Well over three quarters of the households across Native American reservations were building coffins, welding crosses for the cemetery, doing hair, babysitting, catering lunches, all this vibrant economic activity that was going completely ignored. We designed the Lakota Fund based upon what was already there, and what was already happening with the reservation economy. At Calvert, of course, we had to go through our shareholders, which to everyone's surprise, but mine, wanted to put 5 percent of our portfolio in these low income community loan funds! Over night, what had become a radical experiment at Calvert became one of the fastest growing products in the social investment market." Rebecca adds, with some irony, "In an Indian system, if you needed a bank loan, you would qualify for a bank loan. In the system we have, if you *don't* need a bank loan, you qualify because the reward is what you've already accumulated. Everybody knew Indians shared too much and cooperated too much. So, if you can imagine sharing and cooperation as a negative—but the understanding was those were not business principles, that we could not have economic

Shari Berenbach
Calvert Foundation

growth and capitalism with people who got along and shared!" Native American communities have also benefited from tribal gaming and casinos. While such paths to economic self-sufficiency have been criticized (and some tribes were fleeced by now-jailed lobbyist Jack Abramoff), they offered one of the few options available. At least the new revenues from such economic activities have led to improved schools, social services, health care, and many Indian-owned banks, represented by the North American Native Bankers Association. *The Economist* noted this growth in financial services as a positive achievement (Feb. 19, 2005).

Shari Berenbach adds, "Since the Calvert Foundation's launch, we've financed about twenty thousand units of housing. We also provide financing to support small business loans. The foundation has $100 million that has come from more than two thousand investors, all with the objective of seeing those dollars reaching down to the community level, and reaching those people who do not have access to financing, and who really can use that capital to address their pressing social needs."

Another innovative group, the San Francisco-based Rudolf Steiner Foundation (www.rsfoundation.org) helps steer funds from donors and investors into local enterprises to revitalize rural and urban communities worldwide, based on the work of Rudolf Steiner, a European philosopher and educator who founded the Waldorf schools.

The concept of investing in the local community is now a robust global movement. From India to Ethiopia—people are securing small loans to start businesses and ensure their financial independence. However, access to financial capital is still limited despite the fact that big banks are now offering small loans and micro-loans because they have learned that many small-scale borrowers have better payback rates than bigger corporations.

Equal Access undertakes a unique effort to create economic independence by providing *communications* rather than *money* as a path to economic freedom. Money and information are equivalent and often information can revive local communities as well as or better than money. Ronni Goldfarb is executive director of Equal

Access and explains her groundbreaking projects, "We are really targeting those people who have never made a telephone call, who may be largely illiterate, who have very few resources, and actually lack the most basic information that could improve their lives. One of the tools we use is a digital satellite receiver. And the reason for this is that there are many areas in the countries where we work that are not even well penetrated by terrestrial radio. Seventy percent of Afghanistan is rural and after twenty-three years of war the entire country has been devastated. So our first project in Afghanistan was a radio-based teacher-training program. Through that pilot we have trained over 3,500 teachers now benefiting 150,000 students. But since that time because of our success doing that we were asked to place 7,000 satellite receivers into the most poor and remote villages in the country. These receivers are now installed. Our channel broadcasts over four hours of content in Pashtun and Darreh languages and people are getting basic education, teacher training, they are getting rural development information. We also relay BBC World Service programs; the Afghan education program, which has beautiful children's stories and soap operas that talk about coming home to Afghanistan and building a new life. We like to say we are bridging the gap between poverty and opportunity." Equal Access is now distributing 300 receivers in Cambodia.

Ronnie Goldfarb's Friends in Afganistan

Economist Jeffrey Sachs makes a similar case in his *The End of Poverty* (2005), showing that investing in health and education produces large returns and is "the bargain of the century!" Sachs's analysis encouraged the millions of activists worldwide with their campaign "Make Poverty History," led by Bono and other rock stars launched at the 2005 Group of 8 Summit in Scotland. Major progress in ending poverty has been made in Brazil by the administration of President Luiz

Inacio Lula da Silva and his Zero Hunger program. Based on the success of the Bolsa Escola program launched earlier by former governor of Rio de Janeiro, Christovam Buarque, which paid families to keep their children in school, the new Conditional Cash Transfer (CCTs) pays poor families to keep their kids in school, have them vaccinated and their health monitored, and buy needed food and transportation. Mexico has a similar program "Oportunidades," which provides cash transfers to five million Mexican families. *The Economist* called these cash-transfer programs "a better way of helping the poor than many previous social programs" (Sept. 17, 2005). Combined with microcredit for entrepreneurs and assuring land and property titles, as pioneered by Peruvian economist, Hernando de Soto in *The Other Path* (1989), many analysts now agree that the Millennium Development Goals agenda for halving world poverty by 2015 is achievable.

Bill Drayton emphasizes the big picture. "Half the world's operations have gone from pre-modern to entrepreneurial competitive in two and a half decades. This is a big challenge. How do we work out the ways to work together? The more entrepreneurs you have, the more local change makers. The local change makers become role models. Imagine the world where you no longer have just 2 or 3 percent natural leaders and everyone else has defined themselves out of that game. Imagine a world where everyone knows they can do this and they know how to work together to do it. This is the most profound step forward in the evolution of our species." Josh Mailman adds, "The idea of ethical globalization *can't* remain a dream because it has to be an *imperative.* The truth is that everything falls apart the more inequality we have." C. K. Prahalad, the management guru, shows in his *The Fortune at the Bottom of the Pyramid* (2003) that, in fact, the least served marketplace in the world are the two or three billion poorest people. So there are not only ethical, but also business reasons to start to think about how we can meet the needs of the poorest three billion people in the world. Susan Davis agrees that C. K. Prahalad has a very strong hypothesis, in saying that we can actually look at the people on the bottom of the economic pyramid as underserved consumers—as untapped producers. "Any business that's not looking at them as potential partners, producers, or consumers is going to miss the boat." Business is paying attention, including Intel Corporation, whose Shanghai group is competing with Advanced Micro Devices Inc. in India and Taiwan's VIA Technologies to design personal computers costing $200 or less (*Business Week,* June 12, 2006). Ronni Goldfarb reminds us that "information has power, just as much if not more than money. With the right information you can actually make a change in your life." Josh Mailman adds, "Money can do a great job supporting social action, but without social action, the money is useless. So we all have to be social activists in our own way . . . we do what we can, but never think that one person can't change the world 'cause its happening all the time."

> The idea of ethical globalization can't remain a dream because it has to be an imperative. The truth is that everything falls apart the more inequality we have.

Another aspect of community investing concerns monitoring trends—so that negative trends and problem areas can be identified and positive trends can be amplified and rewarded with media coverage. The city of Jacksonville, Florida, led the country in 1983 with its Quality Indicators of Progress Report, pioneered by the late sociologist Marian Chambers and sponsored ever since by the multi-stakeholder Jacksonville Community Council Inc. (JCCI). Today JCCI's annual report still measures Jacksonville's quality of life beyond its economic growth, by educational achievement, health, race relations, air and water quality, nature conservation, public parks, arts and culture, recycling progress, and responsiveness of city government. Sustaining healthy communities requires investment in all these sectors that contribute to overall quality of life. Jacksonville's mayor, Chamber of Commerce, JCCI, and other community organizations that sponsor the report set targets for continuous improvement in all these areas. Local media report progress and where extra efforts are needed. Hundreds of cities around the world have adopted the Jacksonville quality of life indicators as their own benchmarks for community vitality.

Investors have long included municipal bonds in their portfolios. Pension funds and socially responsible mutual funds also include these long-term investments in communities, to build affordable housing and finance the small businesses that create most of their jobs. In 2005 CRAFund Advisors became the first Community Development Investment Manager to earn certification from the Association for Investment Management and Research—making such community investments on par with all other mainstream investment products. Banks are required under the Community Reinvestment Act to make a percentage of their loans available to local borrowers and businesses. In today's globalized financial markets, with some $1.5 trillion of daily currency-trading, local investing is even more vital. In some places like Afghanistan, as we learned, radios are better than money. While hot money roams the planet looking for higher returns, many investors now prefer watching their money work to build their local homegrown economies. Harvard's management guru, Michael Porter, now shows that inner cities are good places to invest. Porter's nonprofit research firm, founded in 1994, released the study *Competitive Inner City* in 2004, which found that retail spending in densely populated areas averages $25 million daily per square mile versus only $3 million in other metro areas. Porter is now researching the top ten cities where their inner-city economies outpace the overall city. Community-oriented investing has now arisen to the same quality and comparable returns as other assets in financial markets—while offering additional social benefits. Today, investors who want to support micro-credit lenders and local communities have a wide range of choices. For more options on these low risk, positive social-investment opportunities visit our Web site www.EthicalMarkets.com.

Community investing now totals more than $4 trillion in the United States and such economically-targeted investments (ETIs) are now a staple asset class. Even financial media are at last picking up on this long-overlooked story. *Business Week*'s May 9, 2005 issue endorsed micro-loan programs as good investments with solid returns, highlighting the Calvert Community Investment Notes we discussed, as well as ACCION Investments (www.accion.org); Blue Orchard Microfinance (www.blueorchard.com); Global Bridge Fund, U.S. Bridge Fund, and Latin American Bridge Fund (also at www.accion.org); MicroVest I, LP, and MPower Investment programs, (both at www.microvestfund.com); Pro Mujer Loan Fund (www.promujer.org); and World Partnership Certificate (www.oikocredit.org). *The Economist* reached similar conclusions in its survey, "The Hidden Wealth of the Poor" (Nov. 5, 2005). Once poor rural people obtain access to loans and banking services so long denied, they prove to be excellent risks and good money managers. Britain's New Economics Foundation's 2005 report, "Basic Bank Accounts," makes the case that banking for all is a universal service obligation (www.neweconomics.org).

The U.S. landscape still has many towns with names reflecting early companies and their founders and benefactors: Alcoa, Tennessee, Kohler, Wisconsin, and Corning, New York, Kennecott, Alaska, and Hershey, Pennsylvania. These towns were built by the successes of those companies and they benefited from the utopian dreams of their founders, who built free playgrounds, zoos, medical clinics, schools, and orphanages. I traced such utopian efforts in *Politics of the Solar Age* from their beginnings in Britain, which led to similar towns such as Bourneville, the home of Cadbury's chocolates. Today, communities cannot rely on such often high-handed benefactors, or even the biggest corporations, as we see the travails of General Motors and Ford—now leading to crises in Detroit and other towns in Michigan. The global auto industry is now dominated by Toyota and China now is a net exporter of cars. Chinese and Indian made cars will range from $2,000 to $6,000 for the cheapest models. Yet as we have shown, communities and local investors and entrepreneurs are succeeding in building and rebuilding homegrown economies and many smaller communities often offer a higher quality of life.

ROUNDTABLE
WALKING THE TALK

Hewson Baltzell, Simran Sethi, and Jean Pogge

Simran Sethi hosted Jean Pogge, senior vice president of Mission Based Deposits at the United State's leading community investment bank, ShoreBank, and Hewson Baltzell, our stakeholder analyst, to look at the social, financial, and environmental impacts of community investment and development.

Simran Sethi: So Jean, what does it mean to be a community-based investment bank?

Jean Pogge: At ShoreBank we define that as being a "triple bottom line" company, where our shareholders and our executive management equally value profitability, community development, and conservation. So we lend in underinvested minority communities and we try to help people understand how to save the planet.

Hewson Baltzell: What kind of wealth creation programs do you have, such as subsidized mortgages, education programs, other programs to help your own employees?

Jean: We provide free single health insurance for all employees, of course, but family coverage costs the employee extra premium money.

So we've scaled that based on income so our lower level employees, many of whom are single moms, pay less than our higher paid employees. We offer seminars for our employees around wealth building issues, for example, how to buy a home, how to get a mortgage, how to repair your credit. These are very popular and they're often taught by our own bankers, so it's sort of a win-win situation for everyone. And we invest in training: on the job training for job skills and a robust tuition reimbursement program. Two years ago we introduced a new program called CEO of Your Career, where each year, every employee gets $500 to pay for any kind of training they want—it doesn't have to be job related. Now, we have said no to golf lessons, but we've said yes to Spanish lessons and other things that enhance your life skills.

Simran: What steps is ShoreBank taking toward environmental stewardship?

Jean: It started with a conservation-based bank, the one out in Ilwaco, Washington, which actually makes loans to small businesses that are trying to create a conservation-based economy. In other words, give back resources as opposed to take out resources from that beautiful countryside out there. Then our board said, Hey, listen, we're not two companies, we're one, so let's bring the conservation mission into our urban areas. That's been more of a struggle, frankly. We try to build awareness among our employees and our customers. We ran contests to reduce paper usage and have reduced it by 40 percent over the last four years. And then we've done special programs on Earth Day and introduced a really interesting program for our single-family home-buyers. We say to them, we'll give you a free Energy Star audit of your home, we'll finance the auditor's recommendations, and if you do the work, then we'll give you a free Energy Star rated refrigerator.

Hewson: Do you actually apply that to your own offices and facilities?

Jean: In our Ilwaco office and in our Portland office we have green buildings. There's absolutely no problem—we built them and they're very wonderful conservation buildings. In our urban areas we're working with older buildings, so it's step by step and it's not quite as good as we'd like, but we've sort of set a goal for ourselves to try and see if we can't get an Energy Star rated building each year.

Hewson: In what ways are your concerns about community and environment built into your lending policies?

Jean: We are a reactive organization, and by regulation, we're not allowed to really run the businesses to which we loan money. So it's tough to impose our standards on them, but what we seek to do is to really help them improve their own standards. So, for example, our conservation bank has a seven-step rating for each of the businesses that we lend to that rates them and their environmental footprint, and we'll lend to anybody on that scale. But once we've made the loan, we have a bank scientist on staff to meet with the borrower and talk with them about ways they can improve their environmental footprint. In urban areas where we work, we recycle old buildings. And then we make sure that our borrowers understand that we're not talking about cosmetic rehab, they have to improve the systems of the buildings so the buildings will stay around for a while and be better for the tenants.

Hewson: So, your major social impact is essentially the employment and the provision of credit where it wouldn't be provided otherwise.

Jean: Absolutely. Keeping neighborhoods intact that might be vacant lots but for this bank. The stories are marvelous. We have rehabbers who start as janitors in their building and they buy one building, then they buy a second, and by the time they have four buildings, they're usually in the business. By the time they have six buildings they have a crew of workers doing the rehab for them and some of them have cashed out in the last few years, $4 or $5 million in their pocket. That's really social impact of the first order, I think!

SIX

Fair Trade

IN THIS CHAPTER WE LOOK AT FAIR TRADE across a range of goods from coffee, chocolate, tea, and bananas to handicrafts, like clothing, household items, and decorative arts. The Fair Trade movement began in earnest at the "Battle of Seattle," in 1999 when thousands of protestors disrupted the World Trade Organization (WTO) meeting there. Founded as the successor to the many rounds of earlier trade talks the WTO began operation in January 1996. The WTO fatally based its rules on obsolete economics from earlier centuries, particularly British economist David Ricardo's ideas of "comparative advantage," endorsed by Adam Smith. The idea was a quite sensible "niche" strategy, where countries, whose workforce, industries, and climate made them advantaged compared with other countries, should build on these natural advantages and trade products with other countries with other special advantages. While economists, especially Adam Smith, lauded the competition of producers, manufacturers, farmers, and workers, all guided by the "invisible hand" of markets, they assumed that each country's economy was largely sovereign, and that capital and finance would remain within a particular country's borders. In an age of electronically accessible 24/7 capital markets, satellites, jet travel, and other globe-girdling technologies, the obsolete economics of Ricardo and Adam Smith—which ignore environmental, social, and cultural costs—are driving today's disastrous global economic warfare, as I described in *Building a Win-Win World* (1996). A report of the United Nations Development Program in July 2006 disputes the economic notion that free trade is a way to reduce poverty (www.undp.org).

In today's world economy, small-scale producers and whole countries can be left out of the WTO bargaining process. When you walk into a coffee shop and order a $5 cappuccino, you might assume that the people who grew the coffee beans were making enough money to also indulge in a cup of java. The sad reality is that the glutted coffee markets have taken a nosedive and the world price, which averaged around $1.20 in the 1980s, slumped in 2002 to 50¢ a pound (the lowest in real terms in the last one hundred years). Since then coffee prices have recovered to between $1.11 and $1.14 per pound—still very low in real terms. This has meant disaster for the estimated twenty-five million global coffee farmers. Fair Trade policies largely remedy this problem, giving producers an opportunity to maintain their traditional lifestyles and earn a living wage. Fair Trade has received a groundswell of support among consumers who look for the many labels that assure them they are not exploiting poor farmers, children, or despoiling natural resources. Worldwide sales of Britain's certified coffee have grown fourfold since 1998 according to *The Economist* (Apr. 1, 2006).

Fair Trade Certified Label

The WTO claims that world trade is a "win-win" situation benefiting all. However, these claims, especially about the poor, have been increasingly challenged—even by former promoters like *The Economist*, which admitted that "the benefits . . . to the world's poorest people have anyway been overstated" (Dec. 10, 2005). The same information technologies that speed global speculation also enable consumers to select Fair Trade products while investors can judge the social, environmental, and ethical performance of companies by new global auditing and manufacturing standards. A study by the teacher's pension fund, TIAA-CREF, found in 2005 that its Trust Index had fallen and that 62 percent of investors surveyed were more cautious due to corporate and political scandals (TIAA-CREF, Winter, 2006).

So let's meet some of the innovators whose efforts have promoted Fair Trade markets around the world. Paul Rice is president and CEO of TransFair U.S.A., which created the standards behind the Fair Trade label on every genuinely fair-traded product. Paul explains, "I used to work in Central America, I lived in Nicaragua for eleven years, working with farmers there, and I got introduced to Fair Trade through that experience. I helped organize Nicaragua's first coffee export co-op in 1990 and had a chance to dramatically increase the incomes of our communities at a time when world coffee prices were very low. So I saw the impact of Fair Trade. I came back to the United States in the mid 1990s and launched TransFair in 1998.

Paul Rice
TransFair USA

"Transfair U.S.A. is the only third party certifying organization for Fair Trade products here in the United States. We certify the companies and the products that meet the international Fair Trade standard. We audit the entire global supply chain. We track those bags of coffee or those boxes of bananas all the way from Central America or South America through the supply chain, through the manufacturer and retailer here in the United States. So when consumers see our Fair Trade certified label they have that assurance that Fair Trade criteria were met, they know that farmers received a fair price. We visit every Fair Trade co-op at least once a year. We visit the farms, we audit the books, and we verify that Fair Trade standards were met at the farm and co-op level. Principally, we're working with small farms that are democratically run co-ops. We then work with the importers who are buying that product and bringing it into the United States. For example, we work with coffee roasters, packagers, and then we track all through to the retail level. We also educate consumers, because at the end of the day, Fair Trade is only successful if we build consumer demand, which is now growing very fast in stores around the country.

"I had a chance to visit some cooperatives in Mexico early this year and took some people from the coffee industry to meet farmers and see the impact that their participation in Fair Trade was making. We stood in the fields one afternoon, talking with a man named Ecedro, who is a fourth generation coffee farmer, and one of eight kids in his generation. They were able to study through the second grade. In their community the norm was that farm kids could study through the second grade, learn how to read and write, and then they were needed in the fields. His generation had never had a chance to study or get an education. Then his community got organized about ten years ago and formed a cooperative. They started to process and market their product directly. About seven years ago, they started to sell to the international Fair Trade market, first to Europe and now here in the United States. And as a result of that dramatically improved income, Don Ecedro's four kids have all finished high school, they have all gone on to college and two of them have graduated from college and come back to the community to work in the co-op. So this is an example of how Fair Trade helps give kids an education. Fair Trade also means empowerment and hope for the future. Fair Trade is really a new form of globalization with a human face. This kind of globalization works for the poor."

Chris Mann is the CEO of the Guayaki Company, producer of teas that are expanding culinary horizons in North America and Europe. Chris has been working with indigenous farmers in Paraguay's rainforests to produce yerba maté, a delicious green tea that promotes energy and health. He explains, "Yerba maté is really an amazing plant, a healthy stimulant and a nutritious energizer. It has twenty-four vitamins and minerals, fifteen amino acids and it is very high in antioxidants. It provides a sustained energy, is good for digestion, and respiratory and circulation systems. It's a beautiful holly, evergreen tree that grows in rainforests. Once a year, we just prune the tree and we are able to take some of the tender leaves and stems, which go through an aging process to produce Guayaki yerba maté. Maté is native to the forest, but has been transplanted from the rainforest to large sunny plantations. This unnatural system doesn't help the forest any, nor the people, but drives down cost and increases yields—typical of large-scale agriculture there, and in the United States for corn, soybeans and other products. We grow maté exclusively within the rainforest where it gets a higher nutritional value in its native environment. Sticking to traditional methods, we can end up with this high quality product. If you go down to South America, especially in Argentina or Paraguay, you will see maté everywhere. It is their symbol of hospitality a daily ritual for some thirty million people.

Chris Mann
Guayaki Company

Harvesting Maté

"Fair Trade is one of the most important global movements happening right now. We now live in a very economic world where we measure things in dollars and cents. People who have gotten pushed to the bottom of that system are the small producers that really hold the wisdom, do the work, and provide food for us. We're very out of touch with all this because it's so far away. The reality is that the purchase decisions we make in the United States have a huge impact on what's happening in the South American rainforest, in Africa, in Asia, all around the world."

Amber Chand, cocreator of The Jerusalem Candle, agrees. "Fair Trade, just look at those words: *fair*, respectful, open, generous; *and trade,* trade is the sacred art of exchange, the most ancient practice that happened since Asia's Silk Road. It is truly the way human beings function—we trade with each other, we barter with each other. Fair Trade is knowing that we must treat people with great respect. As a peacemaker, I must ask myself, How am I in my small and humble way able to find ways to promote peace on the planet?"

Amber describes how Fair Trade commerce can do just that. "Imagine for instance products being built by people in conflict, joint products, which would actually become symbolic of peacemaking. As we know, Palestinians and Israelis are unable to even meet each other. There is this incredible dissention and hatred, between the two peoples. So last year I went to Jerusalem with the Business Council for Peace, a non-profit organization of business women in this country who are committed to supporting women's enterprises in areas of conflict and on whose governing board I serve. We created the Jerusalem Candle of Hope, a product that would be made by both Israeli women and Palestinian women. It is a joint venture which has never, ever been done before. The candle is made by Israeli women who are

Amber Chand
The Jerusalem Candle

Russian immigrants, and it comes with a votive tea light packaged in an embroidered bag made by Palestinian women. So the Jerusalem Candle of Hope is a very powerful symbol of what is possible, when you bring people together. This project is basically supporting one hundred families in Palestine, who now have food on the table. It's supporting twenty-five Israeli women who otherwise would be unemployed, and this is just the beginning of the project."

Some U.S.-based organizations who understand the shortcomings of the WTO model seek partnerships to increase the public's awareness of Fair Trade and sustainability issues, an effort that appears to be working. Fair Trade and whole organic foods markets are growing at 20 percent annually in the United States. Kevin Danaher is the cofounder of Global Exchange, an international human rights organization dedicated to promoting environmental, political, and social justice through these worldwide partnerships. Kevin explains why he and his wife, Medea Benjamin, and their business partner Kirsten Moller, founded this highly successful organization. "The three of us founded Global Exchange in 1988. We are now up to about forty-nine staff. Our budget in 2004 was about $8 million. We have four stores—an on-line store and stores in San Francisco, Berkeley, and Portland—that sell third world crafts from development projects. We do 'reality tours'—a sort of reverse Club Med experience—that take people to other countries to meet real people. We did 142 of those last year. We do corporate campaigning to pressure corporations to change. We do work on reforming the IMF, World Bank, and the WTO. Basically we try to educate people in the United States about our responsibility as the most powerful country in the world. So basically we are an educational organization, trying to sensitize the population to our real responsibility in the world."

Global Exchange's very successful expos of green products—held annually in San Francisco and Washington, D.C., draw many thousands of people. Kevin describes the expos. "The Green Festivals started three years ago. They are a hybrid of a conference and a green economy trade show. We have about four hundred green economy companies, about fifty or sixty really good inspiring speakers. They are weekend events. We have live music both days. We have all organic vegetarian food, organic beer and wine, so it's a party but it's a party with a purpose. We do Washington, D.C., in September, San Francisco in November, and Chicago in April. We had over thirty thousand people come through the door this last Green Festival in San Francisco. Many people came up to us and said thank you for doing this, we really needed this to boost our spirits!"

In many cases, trade unions and international associations of employees have spearheaded fairness in global markets. The United Nation's oldest agency, the International Labor Organization, which also represents employers and governments, released the report of its Commission on Humanizing Globalization, *A Fairer Globalization,* in 2004. The report covered the worldwide efforts to reform obsolete, malfunctioning economic models that ignore social, human, and environmental costs.

Neil Kearney is general secretary of the International Textile, Garment, and Leather Workers Federation, which supports workers by seeking enforceable global rules on labor standards. He believes the international community must ensure that all governments promote and enforce decent work or otherwise forfeit their countries' access to world markets. Neil explains their philosophy. "With the social conditions under which textiles and garments are manufactured these companies know exactly what is happening in the production facilities. The major brand names and the major retailers have representatives there all the time, in the main, checking on quality. You have to be blind not to see what is going on. Many of them have adopted corporate codes of conduct dealing with corporate social responsibility. Today there's something like 10,000 different codes of conduct. Unfortunately maybe 9,700 of those are not really implemented. The labor content in most of the shirts that we buy that are produced in developing countries is no more than ten U.S. cents. Yet, look at the price we pay! We may be paying $20 or up to $50. For some of the fancy brands we may be paying over $100. Actually the consumer gets a poor deal—given the low cost of production, how little labor costs actually are, and, very often, how little the raw materials cost. I sometimes hear the argument that consumers should or would pay *more*. I don't believe that consumers have to pay a cent more to ensure that conditions in this industry across the world are maintained at decent levels."

Neil comments on how WTO rules (driven by the conventional economic source code) create painful disruptions in local industries. "Today about 160 different countries produce textiles and clothing, primarily for export into the markets of only about 30 countries. If China secures 60, 70, 80, or 90 percent of the market, as is occurring under WTO rules, it doesn't leave very much for the other 159 producers. This has dramatic implications for domestic economies, for the social fabric for the countries concerned, and indeed for international security. The WTO needs to urgently examine the affect of trade liberalization on a sector like textiles and clothing to assist what I would describe as emerging and struggling industries just to meet the challenges of the dominant suppliers." The WTO's June 2006 meeting in Hong Kong broke up amid demonstrations by civic groups and farmers protesting further trade liberalization.

Alice Tepper-Marlin, president of Social Accountability International agrees. Her SA8000 label on clothing assures consumers that child labor was not used in production.

As companies from Nike, Reebok, Kathy Lee, and even P. Diddy have found, manufacturing by outsourcing to the cheapest factories in the world often extracts a cost beyond the price tag. Their brand names and stock prices suffered as civic watchdog groups exposed their shortcomings on the Internet and in the mass media.

As China enters a period of labor shortages the game of exploiting workers to provide the world with goods at rock-bottom prices is ending. Producers must offer an ever-more educated Chinese working class higher wages and new benefits. To circumvent this, they move their plants to rural areas or to Vietnam, in a new race to the bottom. Meanwhile, in 2005, Chinese colleges and universities enrolled over fourteen million students—up from 4.3 million in 1999. The WTO with its rigid rules, based on those obsolete free trade economic models, is now experiencing widespread dissention among its own members from developing countries who have recently embraced the Fair Trade model.

We have seen how Fair Trade companies try to link consumers in the United States and Europe with small producers in developing countries to help lift them out of poverty. The good news is that these Fair Trade companies are succeeding. Yet, their success still challenges the two-hundred-year-old free trade models of most economists and the World Trade Organization, which actually preclude many Fair Trade initiatives as unfair competition! Since economic textbooks told us that more trade was good for everyone, the World Bank often advised countries to grow their economies by exporting high-demand commodities to world markets—whether coffee, tea, or computer chips. This often leads to glutting world markets. Economists' recipes for development and GDP-growth urges them to open their borders, reduce tariffs, make their currencies convertible, privatize their main industries, and allow foreign capital to flow in and out freely—often with unhappy results. Many Latin American countries now reject these economic recipes of the Washington Consensus. The new Group of Twenty countries, led by Brazil, China, and India are challenging WTO rules and protectionism in the United States and the European Union. We know how developing countries can lose out. Their weaker, smaller companies and farmers are put out of business and they cannot afford the armies of lawyers and trade representatives needed to negotiate with big rich countries at WTO meetings. We now know that WTO rules are driven by power politics rather than economic realities. The Global Development Institute's David Roodman (also on the Ethical Markets Research Advisory Board) pinpoints the politics of trade in its annual "Ranking the Rich" Index of how twenty-one rich nations' policies on trade, investment, migration, environment, security, and technology help or hurt the poor in developing countries (www.cgdev.org).

Today, currency trading and herds of electronic investors create gigantic waves of global hot money—and battles rage over all these issues. U.S. politicians worry about floods of cheap imports, plant closings, and outsourcing high-tech work—even scientific R&D, formerly the United States' comparative advantage. Yet U.S. consumers still load up on Chinese goods—some $300 billion worth in 2006.

Even well-known free trade economists and the mainstream business media are having second thoughts in the face of U.S. trade and budget deficits, and the weakened dollar. Amidst all this, Fair Trade companies can continue to prosper and good news for small farmers can sometimes come from new companies, such as GrainPro Inc. of Concord, Massachusetts, which developed large, cheap, portable storage containers that keep crops fresh for up to six months—allowing farmers better control over their prices. The new scorecards of wealth, progress, and quality of life are steering policymakers and business toward truly successful kinds of exports, which spread benefits more fairly.

Principles of Sustainable World Trade

- Adherence to all United Nations principles, treaties
- A well-regulated transparent, democratic global financial architecture
- Ending Corruption
- Ending relocation practices based on tax holidays
- Calculating all traded goods and negotiations in full-cost prices
- Truly level playing fields on subsidies
- Correcting GDP per capita based economic growth measures: Rio de Janeiro in Agenda 21 (1992)
- Correcting stock and bond markets evaluations

© Henderson, 2002.

As we've pointed out, most world trade today is subsidized by tax-supported ports, transport, and energy prices, which ignore environmental and social costs. If world trade were to fully account for these huge subsidies we would find that local and regional trade is more efficient. Most countries are capable of producing many of the goods and services they need domestically. Today's perversely subsidized global trade transactions often lead to exchanging identical products, such as the fifteen hundred tons of potatoes that Britain exported to Germany and the same quantity that Germany exported back in 2004. Similarly, ships full of Asian cars, cross U.S. ships full of similar cars in the Pacific Ocean, wasting energy and polluting our planet. A humanized, ecologically sustainable world economy will need to shift away from unnecessary shipping of goods and move toward Fair Trade and services. We must foster trade of ideas, music, culture, and promote cleaner and greener technologies, health and education programs, and peaceful diplomatic agreements on human rights to preserve the Earth's resources. And as we move to communicating more, there will be no more urgent task than redesigning computers and data centers. Today's giant server farms use vast quantities of air-conditioning to cool data processing systems. If they are not made more energy-efficient, global data and communications systems—far from being cleaner and greener—are on track to consume half of all the world's electricity by 2010 (*Wired*, Oct. 2006).

ROUNDTABLE
WALKING THE TALK

Simran Sethi

Bob Stiller

Simran hosted Bob Stiller, president and founder of Green Mountain Coffee Roasters, and our stakeholder analyst Hewson Baltzell to discuss Fair Trade and Green Mountain's strategy, including reasons why 20 percent of its coffee is Fair Trade.

Bob Stiller: Fair Trade is a priority for us because it stands for third party certification, with criteria in the farming of the coffee to assure enough money is going to farmers. We're very passionate about how this makes a big difference in the world and a lot of our employees are very engaged in trying to increase that percentage.

Hewson Baltzell: Aside from Fair Trade issues, your company actually also does a lot of things related to community and to giving back. You have several programs such as the Café Program and your contribution of 5 percent of pretax income to charity. Can you tell us something about that?

Bob: Well, the Café Program is employees really donating their time in the community. We encourage people to be active. We don't really feel it's a choice or that we're sacrificing the success of the company. We feel it's good business practice.

Simran: Have your employees had any opportunities to truly understand what these farmers and producers of coffee go through in their day-to-day production?

Bob: We have sent over 20 percent of our employees to farms over the years. It's something that we've been really working on for fifteen years, to develop sourcing criteria from a social and environmental point of view and for the farmers. We were very happy when Fair Trade came along to do the certification since this adds credibility for the consumer.

Hewson: I realize that you're doing some work regarding global climate change and you've have done some analysis and are trying to offset your emissions by purchasing emissions credits.

Bob: Yes, we have started looking at measuring our emissions—our waste. It has really highlighted some spots where we certainly aren't as strong as we would like. Historically, we have focused more on the social issues than some of the more glaring environmental issues. Doing a CSR (Corporate Social Responsibility) report that covers the whole spectrum has been an eye opener for us and we're really excited to do better and utilize this holistic reporting.

Hewson: On the environmental side that you mentioned, what do you think your biggest impact is and what are you doing to mitigate that impact?

Bob: I think our positive impact has been with the coffee farms.

Hewson: Which you don't own, incidentally, right?

Bob: They are all suppliers—but we got people together for best practices: how to plant shade trees, eliminate the use of various pesticides and herbicides. Domestically, Green Mountain uses cogeneration to recapture heat from its energy systems. We have eliminated some of these emissions, but I don't feel we have as good a scorecard on our footprint. We have done some plantings to offset our emissions as well, but we really would love to have zero impact on the environment.

Simran: What would it take for you to reach 100 percent Fair Trade coffee?

Bob: We're always trying to explain the difference to the consumer. We do buy some Fair Trade coffee and sell it in our normal blends, but most of the coffee we sell as Fair Trade certified. We are trying to balance our growth—because we realize the amount of good we do in the world is really a function of how successful we are.

SEVEN

Women-Owned Businesses

THE WAY THAT WOMEN-OWNED BUSINESSES ARE CHANGING the nation's landscape is one of the most surprising underreported economic stories in the United States. Women have always been acknowledged as leaders in domestic and community activities, and now they're bringing their own models of management and innovation to the marketplace, thus enriching the global economy. Women's economic contributions have been undervalued for decades. The tide is shifting. The number of women-owned businesses in the United States is increasing at nearly twice the national average, accounting for nearly $2.5 trillion in sales. Women-owned and managed companies also employ over nineteen million people. *The Economist*, in its "A Guide to Womenomics," (Apr. 15, 2006) now finds that "women are the most powerful engine of global growth," and that the increased employment of women in developing economies has contributed much more to global growth than China has.

Sharon Hadary of the Center for Women's Business Research based in Washington, D.C., talks about the dynamics of this trend. "Between 1997 and 2004, the number of women-owned businesses increased at twice the rate of all businesses. When you look at growth in the number of women-owned businesses, the expansion in employment and revenue far exceeds the growth in the number. So what that tells us is that these businesses are larger, they are more substantial, and they're making a great contribution to our economy. Women are expanding into all industries as business owners. And, in fact, the fastest growth is in what we might think of as non-traditional industries, for example, engineering, telecommunications and public utilities, construction, agri-business, and wholesale distribution. That's where we see the fastest growth of women-owned businesses." These numbers reflect not only the desire to make a living but also the pull of entrepreneurship, the need for personal autonomy and flexibility, and the desire to break through the glass ceiling that exists in so many corporate environments.

Deborah Sawyer
Environmental Design International

Deborah Sawyer, president and CEO of Environmental Design International (EDI) in Chicago, a licensed civil engineering firm that does land surveys, geotechnical engineering, monitoring, and remediation of contaminated sites for reindustrialization and redevelopment, describes one example of her work. "EDI was hired by the Chicago Department of Environment on behalf of the Chicago Housing Authority to test the soil to make sure it wasn't contaminated before the construction began. We have three different contracts at O'Hare Airport. We have a $3 million dollar civil engineering contract against which we get various task orders for surveying, drainage, etcetera. Then we have an environmental contract where there's still a lot of property acquisition. There's a lot of construction. If there's an area where they're about to do something they have EDI go out and do some environmental sampling to see whether or not the site is clean enough to do whatever construction project they want to do. Or, they'll discover some old tanks or other problems and want those removed or remediated." Deborah reminisces, "In the early days the challenge was financing. We started in 1991. Fourteen years ago there were no financial instruments for small service-based businesses. Banks just did not understand them. There was one year I owed the IRS $1 million and it was all I could think about every day. Are they going to come and chain the doors and put me out of business? A lot of the reason why firms like mine have cash flow problems is because government agencies, many of which are clients, don't pay their bills! And the IRS understood that. When I said *you* go collect this $1 million from this federal agency, they put their tail between their legs and left me with a little payment plan, and everything was fine. In 1995 I was named the minority businessperson of the year for the whole United States! So I won for the state of Illinois, then for the Midwest region and all the regional winners were invited to go to Washington, D.C. My God, they treated my mother like a queen for the whole weekend! That was probably my proudest moment, taking my mom to the White House."

Judy Wicks, another pioneer business-owner, has been previously mentioned in this book for her outstanding success in the building community as founder and president of Philadelphia's White Dog Café. Judy illustrates the broader motivations of women entrepreneurs and stands at the forefront of the movement to meld social values and business objectives. Turning a carry-out muffin shop into a vehicle for championing social causes, as Judy did, is an example of how women's businesses are often geared to supporting communities. Judy said, "I started the White Dog on the first floor of my house

Judy Wicks
The White Dog Café

in January of 1983, so it's been twenty-two years now. And then I gradually expanded it over the years. I feel that the purpose of business is to serve. So the mission of the White Dog Café is very simple, that we want to serve to our full potential in four areas: we serve our customers, we serve our work community (each other as fellow employees), we serve our community, and we serve nature."

Similar motivations have driven women to champion innovations in health care. Dr. Victoria Hale, CEO and founder of the Institute for One World Health, develops drugs to treat "orphan diseases" neglected by commercial drug companies using generic, donated, or otherwise royalty-free drug formulas to treat poor patients, especially in developing countries. Kara J. Trott founded Quantum Health in Columbus, Ohio, in 1999 for similar reasons. Quantum helps patients navigate the complexities of the U.S. healthcare system. "I wanted to create something and make a change in people's lives," Trott told *Business Week* in their story on Quantum's success (Feb. 27, 2006). She started with $400,000 of her own money from her former law career. Ginger Graham launched Amylin Pharmaceuticals, Inc. to serve diabetics like her. "I just think we don't appreciate how hard their lives are." Now Amylin has launched two new drugs to treat this disease and is a publicly traded company whose stock climbed 58 percent in 2005 to $39 a share, according to *Business Week* (Jan. 9, 2006).

Sharon Hadary delves further into her research findings: "We're seeing that women are coming to business more and more with experience that is similar to the men's experience. They're beginning to bring professional and managerial, and even executive experience, to business. But we still see that there are differences in how the women approach their business. Men business owners are much more likely to prefer what we've come to call 'left-brain thinking,' which is characterized by logic and facts and hierarchy. Women business owners are, in fact, more likely than men business owners to use 'right-brain thinking,' which is the thinking that is characterized by values and intuition and relationships. But what's never made it into the common literature, or the common knowledge, is that *women tend to be about half and half.* So, they want to focus on values, values are important to them, they want to build relationships, but they also want the facts and the logic. Women are much more likely when they're making a decision, not only to get advice and information from outside experts, or experts within the business, but they're also more likely to consult with the people who will be affected by the decision." Another myth about women has now been dispelled. Women are just as interested in computers and electronics as men and now account for 50 percent of all technology purchases. Women now head thirty-three million households—up from twenty-one million in 1980.

Women's buying power has soared 63 percent in the past thirty years (even though they still earn only $0.78 for every dollar a man earns). Women also seem to manage investments differently from men. The long list of women who spearheaded the socially responsible investment sector—now at $2.3 trillion in the United States alone—includes an array of impressive entrepreneurs, including:

Alice Tepper-Marlin, President of SA International who founded the Council on Economic Priorities in 1968 and became the "mother" of the social and environmental auditing industry;

Susan Davis, CEO of Capital Missions, who organized the committee of two hundred high-net worth women and founded Investors' Circle (chapter 13);

Geeta Aiyer, founder of Walden Asset Management and now CEO of Boston Common Asset Management;

Joan Bavaria, who founded CERES (the Coalition for Environmentally Responsible Economies) and is now CEO of Trillium Asset Management of Boston;

Amy Domini, author of Socially Responsible Investing (2001) and creator of the Domini Social 400 Index (which has regularly outperformed the Standard & Poor's 500) and is CEO of Domini Social Investments in Boston;

Micheala Walsh, founding president, Women's World Banking, now in over forty countries, and **Rebecca Adamson**, president of First Nation's Development Institute, are pioneers of micro-finance;

Barbara Krumsiek, CEO of the Calvert Group, manages their largest family of socially responsible mutual funds, totaling over $10 billion and launched the Calvert Women's Principles to rate companies' performances in equitable hiring, promotion, and treatment of women at all levels of society worldwide (www.calvert.com);

Alisa Gravitz, featured in chapter 9, president of Co-op America, pioneered the National Green Pages directory of green, ethical businesses and manages the Social Investment Forum, the trade association of the ethical investment industry (www.socialinvest.org);

Tessa Tennant, the first ethical mutual fund manager in Britain, initiated the green investment portfolio at Friends Provident, a large insurance company, and moved to Hong King in the 1990s to found ASRIA (Association of Socially Responsible Investors in Asia), which screens companies in Asia, www.asria.org;

Dr. Judy Henderson, pediatrician (chapter 1), founded the Australian Ethical Investment Fund and now chairs the Global Reporting Initiative (www.globalreporting.org); and

Professor Graciela Chichilnisky, mathematician/economist at Columbia University invented catastrophe bonds used to insure damage from natural disasters, devised the first equitable global emissions trading scheme for the Kyoto Protocol on climate change and invented the International Bank for Environmental Settlements (a "green" IMF) to assure that any "rights to pollute" would be distributed equally to every man, woman, and child on the planet.

Inge Kaul, economist (chapter 1), pioneer of the U.N. Human Development Index. Director of Development Studies, UNDP.

Meanwhile, Wall Street remains a bastion of male privilege as described by *Fortune* in "How Corporate America is Betraying Women" (Jan. 10, 2005). Even *The Economist* wondered why women are "so persistently absent from top corporate jobs" in its special report, "The Conundrum of the Glass Ceiling" (July 23, 2005). Norway is leading the way, ruling in 2006 that all companies must have at least

40 percent women members on their boards, or after a grace period, lose their right to operate and be closed down.

Judy Wicks recounts her personal process saying, "I stepped over a real threshold when I decided *to share* my knowledge and my experience with competitors. When you are focused on actually building a sustainable and just economy, you have to think about the whole economy, about the relationships between the businesses. We have a newsletter that comes out quarterly, *Tales from the White Dog Café,* which announces all of our events that we have in the café. I joke that I use good food to lure innocent customers into social activism. As long as people are coming to eat, why not do more? The first project that I had was actually the most adventurous, in a sense, and that's our International Sister Restaurant Project. I had this dream one day that I walked into a restaurant, instead of asking for a table for two or a table for four, I asked for a table for six billion, please! I am envisioning a world where everyone has a place at the table, everyone has enough to eat, and has a place at the table politically and economically. I had heard about the idea of 'sister cities,' and I thought, Why not a sister restaurant? So I ended up going to Nicaragua, establishing a sister relationship with a restaurant there. That also gave me a marketing tool, in a sense, or a focus for an international program, which we call 'A Table For Six Billion, Please.' After I started doing these international trips, I thought why not have a *local* sister restaurant program right here in Philadelphia? So we started one here, building sister relationships with minority owned restaurants in more ghettoized areas of the city, as a way of building community, developing dialogue, understanding between people in our region." Such successful business examples of "doing well by doing good" have been proliferating for the past twenty-five years. The mainstream media, whose paradigm of "success" still prejudices women, have ignored this development. Yet the Center for Women's Business Research 2006 report, "Women-Owned Firms Doing Business Without Employees," found that the 5.4 million majority women-owned firms in the United States generate $167 billion in sales annually in mostly services, including consulting, agricultural services, construction, transportation, communications, and public utilities. For this and their other studies, visit www.womensbusinessresearch.org.

Paul H. Ray, PhD, coauthor with Dr. Sherry Anderson of *The Cultural Creatives* (2000), comments on these issues. "The Cultural Creatives are a population we discovered in the process of doing fifteen years of research on people's values and lifestyles. Values and lifestyles are key words in the market research trade. Values, in this case, mean values of environmentalism, values that are about women's concerns in life, values of saving the planet, values around alternative healthcare, health in general. Cultural Creatives are also the people who are leading the concerns for women's movement issues, every aspect of concern for women and children around the planet. That wider view shows that environmental concerns exemplify this view that encompasses many new developments in our modern American culture. In the United States, the best estimate we have for Cultural Creative is that there are fifty million adults and in western Europe, eighty to ninety million adults who are Cultural Creatives. In addition, the Cultural Creatives have a spending power that's just above the average for Americans. With fifty million people, that's $1.2 trillion of income after taxes. The primary beliefs of Cultural Creatives are that women's values, women's perspectives really matter. These Cultural Creatives are the people who are creating new culture."

Paul Ray
Coauthor, *The Cultural Creatives*

A 2003 National Women's Business Council Survey on Women's Entrepreneurship in the Twenty-first Century determined that over half the women polled found it difficult to secure capital to start a business. Women's World Banking has been responding to that challenge since 1975. This global network of micro-financing institutions, discussed in chapter 5, has long recognized the key roles low-income women entrepreneurs play within families and communities all over the world. Micro-financing institutions build relationships with local associates and global financial institutions like Deutsche Bank and CitiGroup to provide direct credit services to over fourteen million low-income women from Bangladesh to Morocco. Michaela Walsh, a former stockbroker turned philanthropist, and, cofounder of Women's World Banking tells this inspiring story. "We founded Women's World Banking on the basis of understanding that the environment and the natural systems of our world are not going to be sustained unless people at the very lowest, closest to the earth, were able to help manage that system and to preserve it. By contributing to the financial side of the productive force ultimately will also help to create a way of sustaining the natural environment. If you're going to talk about developing economies, how can you develop an economy that doesn't allow 50 percent of the workforce and the producers in those economies to have access to the same tools of production available to others? We need to figure out a way to adjust the market system and to adjust the financial systems to create equitable and fair rates of interest, service, and cost of services for anyone who wants to be a producer of any kind in their own economy."

Nancy Barry
Women's World Banking

Nancy Barry, president of Women's World Banking follows up, "We have been working, for the last twenty five years, in what is now over fifty countries with fifteen million low-income women. What you can see, whether it's Colombia, or Kenya, or Gujarat, India, is that low-income women have over time, with a microfinance institution or a bank that believes in them, they have built their businesses. Their daughters are now in university and they will tell you, in places as difficult as Colombia, during civil strife and recession, 'My life is better today than it was five years ago, my daughter's life is going to be better than mine.' So, it essentially gives people a step up, which allows them to manifest their dreams."

Michaela Walsh
Women's World Banking

Michaela recalls, "When we first started Women's World Banking, we were talking about how we could get an equal, fair rate of interest for women to borrow money, rather than for a moneylender to be able to go to a financial institution and borrow. Now that evolutionary process has taken us to where we're looking at microfinance, and there is a huge industry out there."

Nancy continues, "What is called the microfinance industry really began about twenty-five years ago. Today there are over sixty million clients served. Our network serves fifteen million of those. There are five hundred million low-income women and their families that still need access to these financial services. Having participated in getting from zero to sixty million, I am much more confident of getting all the way to five hundred million!"

Michaela reminisces: "One of the early people who worked on Women's World Banking was the first African woman to be made a vice president and manager of Barclay's Bank in Africa. She convinced the bank to allow women to open savings accounts for their children. They became the first guarantors, with those accounts, for loans to women in Kenya. Now the business of Women's World Banking in Kenya is one of the largest of the Women's World Banking affiliates."

Nancy adds, "About 25 percent of the fifteen million clients served by the banks and microfinance institutions in our network are men. The reason that the whole microfinance industry movement has migrated to women is first of all, women repay their loans better. What we've learned is poor *women* repay their loans somewhat better than poor *men*, but poor *people* repay their loans much better than average, often richer borrowers. It is also true that when a woman entrepreneur earns one hundred rupees, ninety-two rupees gets back into the mouths, medicine, and schoolbooks of her children. Statistics in India indicate that if a man gets that increased income, only forty rupees gets back into the household economy. So if you are really trying to build economic and social assets for a household, it's more efficient to lend to a woman. About one third of these loans go to women-owned businesses, one third goes to 'mom and pop,' owned firms, and one third, especially in more traditional cultures, actually is taken by the woman and given to a business of the man. The difference is that even in the last case, the woman gets more power. The equilibrium of power in the household begins to shift, and the banker is more confident because the woman will make sure the loan gets repaid." As noted in chapter 5, there are many ways that investors and donors can support all of the microfinance programs we listed.

Susan Davis
Grameen USA

Susan Davis, board chair of Grameen Foundation U.S.A., describes an innovation of the Grameen Phone Company, cofounded by Iqbal Quadir, a visionary Bangladeshi and Josh Mailman. "We documented Grameen Bank's very successful partners, the Grameen Phone Company. They also created a non-profit, Grameen Telecom, to be able to provide villages with phone service. There are now eighty thousand village 'phone ladies,' who provide phone service through a cell phone that's powered through solar energy, given to them through Grameen Solar. We took that case to Uganda. We showed the business model to MTN, a local phone service provider. They weren't necessarily into doing good, but when they saw the revenue model, they found that they could reach an untapped market they hadn't even thought about." Wireless cell phone services are spreading worldwide because they don't require the massive infrastructure needed for landlines. In addition, small-scale businesses and individual entrepreneurs can enter such services with very little investment, often a micro-loan, which soon pays for itself.

Nancy Barry, sums up the case for micro-finance. "Now we can see that it works, and now mainstream actors are seeing that it's working. All we need to do now is keep innovating and make sure that the thing does not get too mechanical. We need to make sure that we stay very close to low-income women who know what they need. Their evolving needs are to cut transaction costs and build domestic capital markets. Then just get out of the way! Because it really is an engine for change."

Starbucks and DELL are among many companies that support women in a different way—by endorsing the Calvert Women's Principles, the code of corporate conduct mentioned earlier. Women continue to make great strides in corporate environments. Female chief executives are running divisions of many Fortune 500 companies (*Fortune*, Oct. 16, 2006).

What accounts for the amazing growth of businesses owned and managed by women? One explanation has to do with potential rewards. *Fortune* noted in 2005 that "a huge pay and promotion gap still yawns." Challenges still face women forty years after sex discrimination became illegal. A recent study by psychology professor Hilary Lips found that the higher a woman's educational level and the pay scale in her field, the wider this earnings gap! In Fortune 500 companies, women now hold half of all managerial and professional positions but make up only 8 percent of executive vice presidents or above. Yet, according to an AFL-CIO survey, about 62 percent of all women in the paid workforce are contributing half or more to their household income. Several studies quoted in *Business Week* report that in many management and leadership roles in companies, women out-perform their male colleagues. Pollsters Celinda Lake and Kellyanne Conway's surveys in *What Women Really Want* (2005) found that women entrepreneurs are among the most philanthropically active businesses; over half give $25,000 or more a year to charity, while 70 percent volunteer in their communities at least once a month. Lake and Conway quote a study of 353 Fortune 500 companies that demonstrates that companies with a higher representation of women in senior management positions financially outperform companies with fewer women at the top. So as long as inequality and glass ceilings persist in corporate America, women will continue to embrace the opportunities and freedoms of entrepreneurship. With the nineteen million jobs women have created so far, that's good news for the economy and society.

ROUNDTABLE
WALKING THE TALK

Alice Tepper-Marlin, Simran Sethi, and Amy Hall

Simran hosted Amy Hall, social consciousness director of the Eileen Fisher Company, designers of women's clothing and our stakeholder analyst, Alice Tepper-Marlin, president of Social Accountability International, to examine the company's social, financial, and environmental impact.

Simran Sethi: Let's talk about Eileen Fisher's community involvement and why they're so committed to this?
Amy Hall: Eileen for a long time has felt very strongly about supporting the community of women both locally, across the country, and internationally. As we are a women's clothing company, Eileen cares about supporting women's causes, particularly women who are disadvantaged—helping them to gain economic empowerment, preventing violence against women, encouraging self-esteem among women, and encouraging health and wellness for women.
Alice Tepper-Marlin: What really is the vision of the company? Can you tell us in practical terms, for staff, how is it different from working at any other company in the rag trade?

Amy: We abide by four basic tenets that are stated in our mission statement: individual health and well-being, joyful atmosphere, social consciousness, and teamwork and collaboration. And those four practices are evident in every location of our company.

Simran: Tell us about the employee wellness program, because that really fosters the kind of environment that supports women whether they're in the office or at home. How does the work you do with yoga and reflexology tie into the bigger picture of the mission?

Amy: Eileen believes that people—whether in the company or elsewhere—need to take care of themselves. Then we're going to avoid having medical or other kinds of problems later on. So we bring practitioners to our offices and our stores and have a wellness benefit where we reimburse employees up to $1,000 a year for wellness-related treatments. In our community grants we fund programs that give access for low-income and disadvantaged women to holistic and integrated therapies.

Alice: The actual sewing, the production in the factory, the assembly of your clothes—that's done in factories, about twenty of them around the world, that Eileen Fisher doesn't own. Neither does it manage them directly. This industry mostly employs women, who in the United States, tend to be immigrants who don't have very good language skills. In China, they come from the most desperately poor, rural areas. How does that fit with Eileen Fisher's vision with the way you like to treat employees?

Amy: About seven years ago, Eileen Fisher made a commitment to understand and develop and improve the workplace conditions in the factories that we were using. We decided to adopt SA8000—Social Accountability 8000—the globally recognized workplace standard by Social Accountability International developed by Alice Tepper-Marlin. This SA8000 sets forth nine fundamental elements, workplace conditions that are minimum standards that the world recognizes for any workplace to operate under. They range from child labor and forced labor all the way to disciplinary practices, working hours, health and safety, those kinds of things. We realized that we couldn't just walk in and hang this standard on the wall, walk away, and expect the factories to be able to follow it, so we partnered with various other organizations—Business for Social Responsibility, Verite, Social Accountability International are the three major ones—to provide varying types of training to the factory managers as well as the workers to help them develop a culture, a workplace culture that supported this standard.

Simran: How is Eileen Fisher managing the environmental impacts of its supply chain?

Amy: The environment, for us, is a very new challenge. I'll say challenge because we're just now beginning a program that will lead us toward a commitment to sustaining the environment though our product and practice. Because of the very nature of the people we hire, the company has actually already done a great deal informally in the way we furnish our stores. We have an organic cotton product in our line that has received a great deal of attention—and so we really have a lot of the pieces in place but this is the first time we're formalizing it.

EIGHT

Renewable Energy

RISING OIL PRICES AND NEW SCIENTIFIC PROOF OF GLOBAL WARMING and rising CO_2 levels from fossil fuel combustion have spurred an increasing interest in alternative forms of energy that reduce our dependence on fossil fuels and lead us into a cleaner, greener future. A study in *Science* (Mar. 2006) predicts that current global temperatures will rise an average of three degrees centigrade by the end of this century. This study also forecasts a rise in sea level by six meters (rather than the one meter previously forecast by the Inter-Governmental Panel on Climate Change), which would flood many of the world's major cities, including Bangkok, London, Miami, and New York. A BBC World Service poll across nineteen countries found 81 percent concerned over the impact of current policies on the environment and climate (July 2006). The flurry of media coverage of global warming is expediting acceptance of viable new policies and technologies that have been blocked so far by the fossil fuel industry and its lobbyists. At last, many businesses are seeing the opportunities and cost-savings in switching to cleaner energy (*Business Week*, July 17, 2006). Key policy changes include removing the huge subsidies the fossil fuels industries and nuclear power have enjoyed, which have prevented renewable technologies from competing; the use of governments' buying power to jumpstart their markets and adjusting tax policies; promoting standards like EPA's Energy Star; and raising CAFE standards on mileage requirements for cars and trucks. All these are favored by 80 percent majorities in the BBC World Service poll.

In this chapter, we look at some of these initiatives—from programs that bring solar energy to rural communities to efforts by companies to reduce greenhouse gases, improve their efficiency, *and* grow their bottom lines. Since the leaders of the Group of 8 powerful industrial countries, including U.S. President George W. Bush, acknowledged at their Summit in Scotland, July 2005, that greenhouse gases (CO_2 and methane) from human activities were contributing to global warming. The heat is on to shift to alternative, renewable, climate-neutral energy systems. Venture capital is now pouring into clean, green technology companies. With $1.6 billion in 2005 and growing by 36 percent annually, the biggest initial public offering (IPO) was China's Suntech Power Holdings $5.5 billion on the Hong Kong stock exchange (*Environmental Finance*, June 2006). Needless to say, the coal, oil, and fossil-fuel-intensive industries are fighting rearguard actions. Meanwhile, U.S. Congressman Bob Inglis, a republican from South Carolina, is sponsoring a bill for a $100 million "Hydrogen Prize" to replace the internal combustion engine (*The Economist*, May 20, 2006).

When we burn a gallon of gasoline in our car, many pollutants, including about five pounds of carbon, are released into the atmosphere. If this carbon were solid, you'd actually see your car's exhaust. Instead, it's released as toxic particles and CO_2—the greenhouse gas that most scientists agree contributes most to global warming. The fossil fuels burned to run cars and trucks, heat homes and businesses, and power factories create toxic air in our cities, contribute to a great deal of respiratory sickness, and amount to about 98 percent of U.S. CO_2 emissions.

A widespread belief, fostered by obsolete economic models, that moving to clean, post-fossil-fuel societies would cost too much, destroy jobs, and lower economic growth still hampers a needed public discussion. In fact, many new studies show that shifting to renewable-based, clean, less-polluting, and healthier societies would spur innovation, economic growth, and create millions of new jobs. Energy/economic expert John A. "Skip" Laitner's "Time to Reassess the Economics of Climate Change," March 23, 2006 (on our Web site www.EthicalMarkets.com) summarizes these opportunities. Skip notes how obsolete economic models overlook the ways that shifting to renewables and increasing energy efficiency would actually increase the productivity of our economy. Many of these technologies pay for themselves in three to five years and have lifetimes of ten to fifteen years. An annotated database shows some 1,500 studies showing overall economic savings from 10 to 40 percent from such technologies. Reducing energy consumption in many sectors could reduce current waste by 5 to 25 percent. These and other efficiencies could save consumers and businesses between $50 and $75 billion in lower bills. The European Union has introduced "externalities pricing," which forces fossil fuel providers to include their full environmental costs. The European Union now leads the world in harnessing wind power, which is now competitive with fossil fuels without subsidies. The International Energy Agency (IEA) forecasts that over $1 trillion will be invested in nonhydro renewable energy technologies worldwide by 2030. In the United States, the 2005 Energy Law handed most of its $80 billion in subsidies to oil, coal, and nuclear power rather than to wind and solar companies. Individual states, led by California, along with the private sector in both the United States and Canada, are acting in many ways to promote renewable energy. Whole Foods, the most successful company of its kind, now on the Fortune 500, bought wind energy credits to cover 100 percent of its projected use in 2006, making it the largest buyer in North America (*USA Today,* Jan. 10, 2006).

Canadian geophysicist Geoffrey Ballard's pioneering development of fuel cell technology is revolutionizing transportation—the largest user of oil—and the way many industries are planning to reduce risks and reverse their toxic impact. Dr. Ballard lives in Vancouver, where he founded The Ballard Power Company. "We're not talking about the billions of dollars for the transfer into hydrogen economy. We're talking about the fundamental research that will give us a fuel cell that is more amenable for competing with internal combustion engines. The fuel cell is extremely simple. It is made up of basically three parts. There is a part that brings in the fuel. Hydrogen goes to one side of the cell, and there is a vehicle for bringing in air to the other side. We use the oxygen in the air, and the hydrogen, and in between there is what we call a membrane electrode assembly that looks like Saran Wrap or a piece of clear plastic. This special piece of plastic allows only the hydrogen nucleus to pass through it, and it kicks off the electrons. The electrons are forced out into an external circuit, and when you push electrons into an external circuit, you have an electrical current. That electrical current is then used to drive electric motors. A fuel cell automobile is really an electric car. When you have a fuel cell engine in a car, you have a way of producing electricity. If you look at the cars in California, for which I happen to have data on, you would find that 4 percent of the cars on the road represent more generating capacity than the entire stationary capacity of the state."

Geoffrey Ballard
Ballard Power Company

Mindy Lubber is executive director of the Boston-based CERES, The Coalition for Environmentally Responsible Economies, founded after the Exxon Valdez oil spill in Alaska by Joan Bavaria, now president of Trillium Asset Management, a socially responsible asset management firm. CERES has pioneered the CERES Principles of Environmental Stewardship and has persuaded over sixty-eight corporations to sign them. CERES also works with many pension funds and asset managers to urge the companies in their portfolios to disclose their plans for reducing their greenhouse gas emissions. Mindy voices her concern about the U.S. industry saying, "Toyota and other Japanese companies are gaining a larger and larger share of the hybrid car market. American companies need to be getting part of that market as well. We want there to be more jobs for America, but at some point, as the price of gas rises, fuel economy standards will eventually get passed. When there are changes to the laws, it will be necessary for the automobile companies to make smaller cars. They need to be doing that now, not later. Not acting is a financial cost to those companies, and shareholders are starting to speak up loud and clear and tell companies to act."

Mindy Lubber
CERES

This kind of multitrillion dollar strong shareholder advocacy did not exist in the 1960s, when a few lone shareholders stood up to management. The 1968 Campaign to Make General Motors Responsible changed all that, as Ralph Nader organized many citizen action groups to seat outside directors on GM's board. As cofounder of New York City's Citizens for Clean Air, I joined this effort—which resulted in GM appointing Rev. Leon Sullivan, the visionary African-American leader to their board. Shareholder advocacy, the focus of the next chapter, for cleaner air and water, less toxic emissions, better workplaces, fairness, social justice, indigenous people's rights, and many other social concerns has become part of annual meetings of U.S. corporations. Shareholder activists have also pushed for clean renewable energy and a shift away from gas-guzzling vehicles to cleaner cars—

from hybrids to electric and flexible fuel vehicles that can run on any liquid fuel and easily convert to hydrogen and fuel cells. The hybrid is a bridge between the current cars and whatever the car of the future will be—quite possibly powered by fuel cells. Toyota is furthest along in experimental fuel cells, which are very closely aligned with their hybrids, such as the best-selling Prius. Honda has also led in hybrids, but only in 2005 did Ford announce its own hybrid, an SUV. Hybrids are a considerable step in the direction of lower fuel consumption, cleaner air, and pointing in the direction of future automobiles worldwide. Geoffrey Ballard predicts, "It will get to the point, very, very quickly, as the fuel cell engine begins to replace the internal combustion engine, that you start plugging your car into your fishing camp. Instead of running land-based power lines, you can start tying into the grid another way. You can have a cabin in the woods. You'll just drive your car up to it and plug it in, and it'll light up the house."

Kyoto has passed. With so much going on in global warming, given the severity of the problem, regulating that problem is good for business.

The Kyoto Protocol, which sets targets for greenhouse gas emissions, came into force in 2005, when Russia, along with 132 other countries, ratified the treaty. The United States and Australia have yet to sign it. Russia nets about 17 percent of global greenhouse gases; however, the United States, with 5 percent of the world's population, produces 25 percent of all CO_2 emissions (*Time*, Mar. 26, 2006). Yet many major U.S. and European corporations, including Dow Chemical, GE, Britain's BP, and others, are now committed to reducing their greenhouse gas emissions—because this often saves money. Even Wal-Mart has joined in, after finding how much money can be saved with more efficient use of technologies and energy. Mindy Lubber adds, "Kyoto has passed. With so much going on in global warming, given the severity of the problem, regulating that problem is good for business." At a senate hearing in April 2006, Duke Energy, Wal-Mart, General Electric, and Shell Oil were among a group of companies urging Congress to impose mandatory caps on carbon emissions.

Companies haven't looked at efficiency because they have been operating with a malfunctioning economic source code. By treating social and environmental costs as externalities that could be omitted from company accounts, they did not see the waste going up their smokestacks, into human lungs, and people's water supplies until the huge backlash of the consumer and environmental movements forced them to clean up. Ironically, the United States sponsored most of the original studies on global warming back in the 1970s and is still a leader in climate research. Today, many American companies are reaping these greater efficiencies and savings from cleaner energy sources such as wind and solar power. Michael Marvin, former president of the Business Council for Sustainable Energy in Washington, D.C., offers some background. "The Council was created in 1992, immediately after the Earth Summit in Rio de Janeiro by energy executives who didn't accept the theory that you had to choose between economic growth and environmental protection. Until 2000, the United States was willing to stand next to its international allies, and deal with environmental issues, regardless of which party was in charge of Congress or the White House. In the past five years, as we walked away from some of these negotiations, specifically the U.N. Framework Convention on Climate Change (as the Kyoto Protocol is called), we see U.S. leadership being challenged, and we see some economic challenges that will come from that." The popular movie *Who Killed the Electric Car?* tells the story of industry rearguard actions to prevent the shift to clean energy. Meanwhile, sales of snazzy commuter bikes were up 25 percent in 2005 and some 50 percent in 2006 (*Business Week*, June 12, 2006).

Geoffrey Ballard looks at our global future, "I don't think we can continue to burn fossil fuels. I think that the big issue that we have today is the fact that China and India and Malaysia have come on stream, wanting the same standard of living as we have. If they choose to use gasoline and internal combustion engines, Mother Earth is going to have difficulty handling all the pollution it involves. Only about 12 percent of the world's population has access to transportation. The other 88 percent is still out there without it, but that 88 percent of the population do have very good communications right now. They're watching American sitcoms. They're seeing the way we live in Western Europe and in North America. They're human. What they want is the same opportunities that we have."

Mindy Lubber sees a similar picture. "As there are more industries creating emissions we could be facing a yet bigger problem. On the other hand, I'm heartened by the changes across the world, in many industrialized nations, in Europe and even in China. There is a growing understanding that we need to create less pollution, not more, that we need to build cars that are more efficient, that we need to build more energy efficient buildings. We see the ethic across Europe that small is OK, that bigger cars aren't the only answer to happiness, as we seem to think here in the United States. We are making progress. Acting now is a *must* because if we wait five years it will be twenty times harder to bring the problem under control. Not acting will cost us money and is bad for business so we've got to move."

Michael Marvin looks at the growing investment in renewable energy. "I think on the renewable energy side right now, looking at just the cost curve and market penetration, you have to look at wind energy. Wind has gone from just a fraction of a percentage in the state of California to now generating more than ten billion kilowatt hours in the United States. The United States was a world leader in wind energy generation through the 1980s and through the early 1990s. And now we're the third largest user of wind energy in the world." Global wind-generating capacity increased by 24 percent in 2005 to 59,100 megawatts and is growing by some 29 percent annually, according to Lester Brown, *Earth Policy News*, June 28, 2006 (www.earth-policy.org).

In terms of solar, we are beginning to see advances made in Germany, China, and Japan, in particular, and global sales are now at $11 billion—up from $7 billion in 2004. In fact, global demand for silicon for the solar cells now exceed supply and suppliers, including Hemlock Semiconductor Corporation of Michigan which is building a new $400 million plant to expand production by 50 percent (*Business Week*, Feb. 6, 2006). Of the ten largest wind turbine manufacturers, one of them is in the United States: General Electric. In 2006, GE, under its new CEO, Jeffrey Immelt, has made a major investment commitment to alternatives, including wind turbines and solar (chapter 4). Michael Marvin notes, "It's an encouraging future. The question is how we can bring those technologies along in ways that underscore the message that they promote economic growth that is sustainable for generations to come. When you're looking at sustainable energy companies from an investment perspective you really probably could look at them in the same way that you might look at a biotech company where potential is tremendous."

Susan Davis
Capital Missions Company

Susan Davis, president of Capital Missions Company, in Wisconsin, an early pioneer of socially responsible venture capital, also created the online computer simulation demonstrating how SRI investments beat many benchmarks on Wall Street (www.capitalmissions.com). "Actually, energy and social investing are a perfect fit now because social investments have proved that they match or outperform financial benchmarks and energy is center stage because of a confluence of the increasing importance of global warming and because of homeland security. There have been three major new funds, by some of the most sophisticated investors in the world, started in the multi-hundred dollar range, particularly in solar, but generally in clean energy." Susan said. In 2006 a flurry of new clean-tech stock indexes appeared, including Ardour Global Index (AGI),

Next Generation Index (NGEX), Nasdaq's Clean Edge US Index, the Cleantech Venture Index, and others (*Environmental Finance*, June 2006).

Innovations such as "smart meters," soon to be mandatory across Europe, reward consumers for tailoring their energy use to reduce overall peak demand. Others are the host of new companies offering "green tags" and certificates that allow people to reduce overall fossil energy use by buying green renewable energy even when it's not available locally. For example, TerraPass, Inc. sells green tags for up to $80 per year to guilt-ridden SUV drivers. Starbucks pledged to buy 20 percent of its annual electricity from renewable sources, through 3phasesEnergy, the intermediary that connects Starbucks, IBM, and Johnson & Johnson to these sources. NativeEnergy.com sells certificates to consumers so that their weddings and events can be climate neutral and their cost, calculated on the mileage of travel for guests, is passed on to help build wind generators on Native American reservations. I tried giving these certificates to our sixty-five guests at a recent family gathering. They were intrigued and delighted.

Amory Lovins
Rocky Mountain Institute

The Rocky Mountain Institute has led in the shift to eco-efficiency and renewable energy since the 1970s and provided a working model of sustainability in its headquarters building in Colorado and in its organizational strategy. Amory Lovins, with Hunter Lovins, cofounded the Rocky Mountain Institute, which has maintained for decades that eliminating energy waste can save billions of dollars. They promoted the concept that conservation is just the same as a new supply—just as useful as striking a new oil well or building another power plant—and have effectively changed the thinking of policymakers in many countries. Amory envisioned a new kind of power grid in which homes and businesses can generate their own electricity and savings. Amory explains, "For the first century of the electricity business, power plants were more expensive and less reliable than the wires—the grid—that moves the electricity from the power plants to the customers. Now new power plants are cheaper than the grid and a lot more reliable. Ninety-eight or 99 percent of power failures originate in the grid, about 95 percent in distribution. So now, if you want to deliver affordable and reliable power to customers, you ought to make it nearer to the customers. This is called distributed generation. The biggest benefits come from the financial economics. For example, it's much less risky to build a *small* power plant *quickly* than a *big* one *slowly*. We can use the tools of portfolio management to quantify what that's worth—typically, close to a factor of three increase in value. Our headquarters building actually exemplifies that. We *make* five or six times as much energy with solar cells as the household *uses*. Most of it we use to run the office. The rest we sell back to the utility at the same price. It's kind of fun to get a call from your utility saying that we haven't had a net bill for you for months. We're using about $5 in electricity per month for about 4,000 square feet, 372 square meters. This is about one tenth of the U.S. norm and we actually know how to save two thirds of what's left but haven't bothered yet. It can go below minus forty degrees out there, and you can get frost any day of the year, but the building has no conventional heating system. It was actually cheaper to build that way. It saves 99 percent of the normal amount of energy for space heating and water heating. It saves 90

Rocky Mountain Institute, Colorado

percent of the normal amount of electricity and half the water. The extra cost of all those savings paid for itself in ten months using 1983 technology."

Amory has been showing the world how superefficient products can also reduce pollution and lower greenhouse gas emissions. "The heat comes from the windows, which have about twice the normal area. They have special films that let in light without letting heat out. They have krypton gas between the panes, which insulates twice as well as air. We also get some heat from the people, windows, lights, and appliances. The real trick is not to lose the heat in the first place, then you don't need much heat to balance that loss. The last percent or so of heat lost can be filled by using a wood stove if you need it. We're collecting energy in the building's central atrium in five different ways. Most obviously, we're collecting heat, which is stored in all the soil, the concrete, the soil under the concrete, the walls, the plaster, the fifty tons of oak beams and so on. There is so much stored heat actually that if we just coasted during a total eclipse—even in January with no heat input—we would lose about a half a Celsius degree per day. Putting the right technologies together in a rather advanced way gets very large resource savings and typically costs less to build and works better."

Amory pioneered the Hypercar design, which began challenging Detroit to do better by showing it was possible for cars to do one hundred miles to the gallon. Now even Washington policy wonks are on the bandwagon. Fareed Zakaria wrote a breathless article in *Newsweek*, April 10, 2005, entitled, "Imagine 500 Miles Per Gallon." Even oil companies now urge us to use renewables. BP now uses the slogan "Beyond Petroleum" to tout its own renewable investments, and Chevron ads tell us "The World Consumes Two Barrels of Oil for Every One Discovered." China has now enacted laws to increase its renewable energy production from 1 percent to 10 percent by 2010 and its auto emission standards are higher than those in the United States. By 2006 the craze for biofuels raised new issues. Should food crops like corn and soybeans be grown to fuel wasteful cars while so many are hungry? Such crops also use lots of energy and water to grow. Only biofuels from cellulose like waste wood chips, sugar cane stalks, and switchgrass can avoid such conflicts.

Often overlooked, the power of the sea can also be harnessed to provide energy. For example, using undersea turbines developed by Blue Energy Canada (in Vancouver) proposes to harness fast-running tides in many straits around the world, such as Puget Sound, Washington. Sea turbines mimic wind turbines, capturing flowing water, which contains eight times more energy than wind. Turbines can be scaled down to harness rivers and serve rural areas. A key strategy for sustainability is to provide on-site power in rural areas instead of depending on costly, inefficient transmission lines from big centralized power plants.

Blue Energy Canada's Tidal Fence

Bob Freling heads the Solar Electric Light Fund (SELF) and brings electricity to such rural places that may never have central generators or power lines, but have an abundance of solar energy to harness. "SELF is a non-profit organization based here in Washington, D.C. Our mission is to bring solar electricity to the two billion people in the world that still do not have access to conventional power. We operate primarily in the developing world, in Africa, Asia, Latin America, and the South Pacific. Our focus is providing sustainable energy solutions to rural communities that have little hope of being connected to an electrical grid any time in the foreseeable future. People that live without electricity typically use

Bob Freling
Solar Electric Light Fund

kerosene as a source of lighting. Kerosene is dangerous, polluting, and very, very bad for their health. People who live with kerosene are smoking the equivalent of two packs of cigarettes a day. Respiratory illnesses throughout the developing world are very, very common. The BBC recently came out with a report talking about the large numbers of people that die every year from indoor air pollution. Bringing clean electricity, clean lights into the home, really helps to prevent this problem, so there is a very immediate health benefit in terms of bringing light into the home. Beyond that, it gives people the chance to engage in productive activities after the sun goes down. In these villages where there is no power, typically after six o'clock in the evening, the night curtain falls. People retreat into their homes that are lit dimly by kerosene candles, and there's virtually nothing they can do productively. Bringing lights into the homes allow parents to engage in productive activities. More importantly it allows children to read and study at night—a very profound implication in terms of the education of children."

Bob describes another SELF initiative, "We did a project two years ago in Brazil. We worked with the Amazon Association and a group of Indians in a very pristine, remote part of the Amazon rain forest. We brought solar electricity to these people which is now being used for water pumping, schools, the health clinic, and a communication satellite dish is delivering high speed Internet access to this community. This has actually been a huge benefit to them in terms of sustainable development. They are able to have a higher quality of life, with access to better education, better health care, and still stay in the rain forest. This is an experiment of the Amazon Association to show how these Indians can live sustainably in the rain forest without destroying it."

So, thirty-three years after OPEC first quadrupled oil prices in 1973, the United States finally got more serious about clean renewable energy. Even though prices have topped $70 a barrel, oil is still less costly in constant dollars than it was in 1973—the drive for less dependence on imported oil is now in high gear. Visionary entrepreneurs, small companies, and venture capitalists are making it happen. The Solar Energy Industries Association reports that solar power alone is now a $5 billion-a-year global industry—expanding at 40 percent annually.

Most sustainable experts agree that if the current perverse subsidies that encourage unsustainability were repealed—solar and renewables could compete handily. I served in the late 1970s as an advisor to the U.S. Office of Technology Assessment with former NASA administrator, James Fletcher. He told us that if such perverse subsidies had instead been given to renewable energy and efficiency, the United States would have already become this kind of sustainable economy.

Wind energy is now the fastest growing form of electricity generation. In Europe, wind provides more than 25 percent of the electricity in some regions. While wind energy generates much less of U.S. electricity, our heartland—great plains—are blessed with a virtual OPEC of untapped wind energy, enough to power the electrical needs of the whole country. Texas and California are leading, while Midwest states and their farmers are pushing biofuels. Both wind and biofuels, as well as solar, are brightening farmers' futures.

Cutting huge subsidies to fossil fuels is crucial. Farmers are becoming energy suppliers. Big energy companies are getting into the act. The $9 billion-a-year FPL Group operates 40 percent of U.S. wind farms in fifteen states. General Electric sees some $2 billion in annual sales and servicing of its wind generators. The big story is that costs are now often competitive with fossil fuels—as well as nuclear, once the social costs of handling radioactive wastes and huge government subsidies are taken into account. Energy efficiency is now paying big dividends. Innovation in solar, wind, hybrid cars, fuel cells, and other renewable energy technologies in the private sector are all driving this shift. A smart national grid will help—but so far, government is behind the curve. A study for the Climate Stewardship Act introduced into Congress by senators Joe Lieberman of Connecticut and John McCain of Arizona in 2005 estimates that this legislation would create more than eight hundred thousand new jobs in the

United States with the largest percentages in Michigan, Minnesota, Nevada, New York, Ohio, South Dakota, and Wisconsin. The mayors of more than two hundred cities have signed the U.S. Mayors Climate Protection Agreement pledging, among other things, that they will meet the Kyoto goals in their cities by 2012 (*Time*, Mar. 26, 2006). We now know that shifting to renewable energy can reduce greenhouse gases and global warming, grow our economy, and produce thousands of new jobs, while improving our health and the environment. Even though oil prices dipped in late-2006, due to seasonal factors and the unwinding of many speculator's futures contracts, most experts agree that global oil production is peaking and prices will stay in their current range—unless global events and crises push them even higher.

Lastly, let's not forget human energy amplified by elegant designs: from the new bicycles to the treadle pumps and other rural technology innovations now used by millions of small farmers and produced by such firms as India's International Development Enterprises.

ROUNDTABLE
WALKING THE TALK

Simran Sethi with Mark Farber

Simran hosted Mark Farber, vice president and cofounder of Evergreen Solar Corporation, and our stakeholder analyst Hewson Baltzell to examine Evergreen Solar, a creator and manufacturer of photovoltaic modules that are the engines of solar electric systems used in remote power and emerging grid connected markets.

Simran Sethi: How does Evergreen use solar energy to benefit its shareholders and customers as well as the environment and the communities in which it operates?
Mark Farber: We serve primarily two markets—so-called off-grid and on-grid. Off-grid refers to places like wireless telecommunications where people don't have access to electricity. On-grid refers to homes and office buildings where we live and work, where solar provides supplemental electricity for those who want some clean electricity.
Hewson Baltzell: On the one hand you have very good environmental attributes, of photovoltaic in general, solar in general. But what about *making it*—what kind of issues come up in the manufacturing process?

Mark: Evergreen's proprietary technology is aimed at lowering the cost of making solar cells and solar panels. Three of the ways that we do that are to use *less* silicon, the key ingredient in making solar cells, using *less* acids in the manufacturing of the product and therefore *less* effluents to clean up, and third, using *less* energy.

Hewson: Mark, you recently announced that you'll have a joint venture with a company in Germany and you'll be building a plant with three or four hundred employees. Can you tell us about that and are there any environmental aspects to that facility?

Mark: We are expanding dramatically in Germany. This will approximately triple our capacity. Germany, as you know, is even more environmentally oriented than the United States and, therefore, energy and environmental issues will be a key component in designing and operating the factory.

Simran: Can you tell us a little bit about the corporate culture in which people work? This industry is a basic manufacturing industry, yet it's one about which people are very passionate.

Mark: In fact, we are a manufacturing company, 70 percent of our employees are production operators, but they really enjoy the fact that we are making, not just another widget, but solar panels that produce clean electricity for people around the world.

Hewson: Much of the solar industry is subsidized right now, both in the United States—in some states—and in many parts of Europe. Where will that go in the future and how does that affect your business? The whole question of subsidies is crucial to the future and to the transition to sustainable societies based on renewable forms of energy.

Mark: It's true that many existing subsidies to fossil fuels and nuclear energy have created an unfavorably tilted playing field for solar. There are a number of markets that are natural, that are not subsidized, primarily the off-grid markets, but there are policies and programs by local and national governments to support these markets for solar and renewables. They are very instrumental in driving market growth.

Germany and Japan are the two largest markets today because of these programs.

Hewson: I'm interested in how solar and renewables can particularly benefit the poor—those two billion people with incomes of less than $2 a day—at the bottom of the world's income pyramid. What do you think about the opportunities for the bottom of the pyramid? That's where people are most likely to be off of the grid, where some of the poorest people in the world live and who don't have access to energy.

Mark: It is remarkable that a third of the people on the planet—two billion people—do not have access today to electricity. That is a key market for solar. It's obviously challenging to get solar panels to these customers, but it is a key opportunity. In fact, our very first sale was providing a solar panel and two light bulbs to about 120 one-room school houses in rural Bolivia. The world *needs* solar!

Simran: And once those panels are there, what's the life cycle of the panel and what kind of take-back policies, like those that are in effect in Europe, do you have in place to retrieve them?

Mark: Solar panels do last a long time. We offer a twenty-five year warranty on our products and they need very little maintenance. We don't expect a problem at this point, since most of the materials we use are very benign.

NINE

Shareholder Activism

In this chapter, we take a deeper look at the rise of concerned shareholders who are investing in the stock market and mutual funds for financial returns on their investment *and* for the betterment of people and the planet. As individual owners buying socially responsible mutual funds, shareholder activists are promoting the shift to sustainability. Many also demand that their pension plans use such broader criteria as the new integrated triple bottom line accounting standards. Such changes affect a corporation's accountability to all its stakeholders, rather than only shareholders. Thousands of active shareholders, together with hundreds of asset managers of socially responsible mutual funds and pension plan trustees, are driving the redefinition of success in the global marketplace. Every year, through these activist shareholders, thousands of proposals are brought to corporate annual meetings, covering a wide range of social and environmental concerns. In the 2006 annual meetings, shareholders filed proxies on over three hundred social issues. Top concerns were the environment, equal opportunity, and political contributions (*Business Week,* Apr. 17, 2006). The Interfaith Center on Corporate Responsibility (ICCR), representing some $30 billion of church portfolios and pension plans, reports on these companies and the shareholder resolutions are presented at their annual meetings at www.iccr.org.

In 1970, University of Chicago economist Milton Friedman famously proclaimed there is only one social responsibility of business—to engage in activities designed to increase its profits or to maximize shareholder value. While some economists still promote this single bottom line view, a lot has changed. The past thirty years has seen the evolution of more ethical twenty-first century capital markets—largely due to the democratization, transparent information flows, and the rise of civic organizations in the global economy. These changes would surely have surprised Adam Smith, the father of capitalism, who wrote his famous book *The Wealth of Nations*, published in Britain in 1776. Today, shareholder activists use their status as partial owners of companies to push corporate management and boards of directors toward greater social and environmental responsibility. After the corporate crime wave uncovered after the Enron debacle, the public largely supports greater oversight of corporations—making corporate crime-fighters, such as New York's Eliot Spitzer, a popular hero. The Sarbanes-Oxley Law, passed in 2002, cracked down on many corporate abuses, such as overstating earnings, which led to some twelve hundred companies having to re-state. Huge salaries of CEOs rose from 33 times average workers' pay in 1980 to 104 times in 2004—outraging shareholders, workers, and voters alike (*Fortune*, July 10, 2006).

Veteran shareholder activist Alisa Gravitz, executive director of Co-op America (chapter 2), has an MBA from Harvard and is an expert in corporate organization and securities law. "This is a really good time if you're concerned about these social and environmental issues to become an activist shareholder. When you own a stock or when you're part of a mutual fund or pension beneficiary you are literally the owner of a company. You have both the right and the responsibility to tell management how you want the business of the company conducted."

Tim Smith (chapter 1), another long-time shareholder activist who founded the Interfaith Center on Corporate Responsibility (ICCR), is now vice president of the Boston-based Walden Asset Management and president of the Social Investment Forum. Tim began as a minister who was deeply interested in Christian teachings on justice and the poor. ICCR continues to monitor corporate social performance, while Tim moved on to head the group of firms that manage portfolios of companies screened by social, environmental, and ethical auditors. Tim reminisces, "For almost thirty-five years now, investors have been engaging companies using shareholder resolutions, dialogue, encouraging companies to act as good corporate citizens. For example, a religious group, a foundation, a mutual fund, a money manager like Walden can sponsor a shareholder resolution. It gets voted on by every shareowner at the stockholders meeting. It's got a tremendous kind of ripple affect as a strategy. We've seen over the years, companies paying more and more attention to the input from concerned investors. Companies understand that consumers, employees, students who are going to become employees of the future, their various stakeholders are also concerned how businesses act. They want them also to be good corporate citizens. So increasingly, you'll see companies by the hundreds doing corporate social responsibility reports, or corporate citizenship reports—telling their story about how they're trying to be a good corporate citizen. And they're doing this not because they simply believe they have a social obligation, they know it's good business!"

Shareholder activism played a part in dismantling apartheid in South Africa through divestment, in phasing out polystyrene containers at McDonald's, in increasing the amount of recycled and alternative fiber in the paper sold at Staples, and in persuading Home Depot to carry sustainably-harvested wood. Citizens *can* make a difference. Linda Crompton, until recently, was CEO and is now an advisor to IRRC, a Washington, D.C.-based research group. She explains, "IRRC stands for Investor Responsibility Research Center. Sometimes people confuse IRRC with being a shareholder activist firm. Not so, IRRC is simply the provider to a whole host of people who are interested in the information, all of the data that concerns publicly traded corporations. IRRC is used by institutional investors to help with their decision-making and by pension fund administrators, state treasurers, all of whom are looking for an unbiased source of data. The other half of the organization is concerned with voting proxies. If you are responsible for ensuring that shares get voted as a fund manager or a mutual fund—you can either do that yourself, which would mean that you would have in-house staff to do it or you would outsource it to groups like IRRC. They can give IRRC a very blanket instruction: that all their votes will go with management (a number of years ago that was very much the common thing). Or, now, with the rise and interest in voting, clients can say these are the priorities that we consider important and IRRC will work with them, to construct a kind of voting framework, depending on the issue that's coming.

Linda Crompton
Investor Responsibility Research Center

"The company was formed shortly after the decision was made to allow shareholders, in effect, to have a voice in terms of what kinds of issues could be voted on. That followed protests over the Vietnam War. The first area of research or investigation that IRRC did was in South Africa on the issue of apartheid. There were a number of large foundations that were concerned about getting a source of impartial and objective information to help them with their investments in South Africa. So they all put money together and formed IRRC."

The Calvert Group was the leading mutual fund to divest from companies doing business in South Africa. I remember that day in the 1980s when South Africa renounced apartheid. I was attending a meeting of Calvert's Advisory Council and all three major TV networks arrived to interview Calvert executives on the role they played.

Gary Brouse, an innovative shareholder activist, a Native American, and director of Corporate Governance at ICCR's New York headquarters, shares an office with church leaders of the National Council of Churches of Christ in the United States. Gary explains, "I think that the more severe cases—like South Africa where there was a stalemate with the companies—led to our organizations. The companies were failing to listen to these concerns. People said that the only way to communicate that message is to begin divesting in companies that wouldn't listen. I can tell you that since the divestment in South Africa, it has always been on the radar of the corporations to be concerned. What you see today is more social criteria being used for investments. So corporations are looking at these new social criteria to make sure there are policies and practices in place so they don't get screened out of some of these portfolios. It is one of their concerns. That's one of the benefits that divestment in South Africa has created for us. Today, and in the future, addressing corporations through shareholder activism is taken much more seriously because of the South Africa victory over apartheid, which U.S. shareholders supported.

Gary Brouse
Interfaith Center on Corporate Responsibility

"If you look at the crises that have been happening in the corporations from the Valdez spill with Exxon; WorldCom; Enron; racial discrimination cases; to the war in Iraq and questions about corporate contracts . . . organizations like ICCR and the shareholder movements have been addressing these issues *before* they have become a crisis. That's why it's important that accountability and transparency exist in these companies. It's one of our main objectives. Whether we've been involved in a written communication with the company or a dialogue with the company or filing a shareholder resolution, we've been communicating with the corporations themselves. This way, the shareholders, the employees, consumers, and the public know what the company is doing and what it's supposed to be doing."

Gary Brouse adds, "It's the responsibility of everybody, not just the stockholders or the corporate executives, but stakeholders, too. Everybody has some kind of investment in a company—whether they're paying taxes or whether that corporation is in their community and it's affecting their community, whether providing jobs or polluting the waterways. Everyone has a say, they're stakeholders in the corporation and can have some kind of influence with corporations. Too many times people try to resolve the issues by looking just at one aspect of our society, whether it is the political system, government, or the laws rather than addressing some of our issues directly with the corporations themselves."

Gary also spearheaded efforts to make companies aware of how their advertising and media buys often led to demeaning stereotypes of Native Americans and often to the outright stealing of tribal names and icons.

Alisa offers other examples of success. "Socially responsible investors got companies like JC Penney to not carry thermometers made out of mercury, which is a terrible risk in your home, and for the people that make the thermometers. Social investors are weighing in on HIV/AIDS, and are getting companies like Chevron/Texaco and Exxon/Mobil to change their policies in the countries in which they operate to help people with the terrible AIDS problems. Social investors are getting companies to pay attention to climate change—companies like American Electric Power and Synergy, because of the pressure that social investors are bringing to bear."

Sometimes shareholder activists join with students, NGOs, and unions in direct campaigns such as the one against Coca-Cola's operations in their Colombia bottling plant, where a workers' strike action resulted in four deaths. The International Labor Rights Fund joined with the United Steel Workers and sued Coca-Cola, while many students at New York University, the University of Michigan, Carleton College, Bard College, and others forced Coke machines off campuses (*Business Week*, Jan. 23, 2006).

In Canada, several groups, including the Coalition to Oppose the Arms Trade, Physicians for a Smoke-Free Canada, and InterPares (a justice group) challenged Canada's Public Pension Investment Board (CPPIB), which holds assets of more than $90 billion, for investing in Lockheed Martin, Raytheon, Northrup Grumman, Halliburton, and other military contractors, as well as in tobacco companies (*Toronto Star*, Dec. 22, 2005).

The United States' largest pension fund is TIAA-CREF, which advertises "Financial Services for the Greater Good." A coalition of civic groups picketed TIAA-CREF's 2005 annual meeting demanding to know why their portfolios included Phillip Morris/Altria with its Marlboro cigarette brand, Wal-Mart, Unocal, Coca-Cola, Costco, and other companies that do not meet the qualifications of a socially responsible business.

Labor unions, including AFCSME and shareholder activist leaders like Rich Ferlauto (chapter 2), are increasingly employing similar strategies by using their pension fund power to augment the power of their capital. AFL-CIO leaders are expanding their own financial services to include ULLICO, Inc., the labor-owned insurance company, Amalgamated Bank, and Union Privilege, which has thirty-one million labor union credit card holders (*Business Week*, Mar. 28, 2005). Although membership in U.S. unions has slid to 13 percent, their coordinated efforts still wield influence, and the union movement is far from obsolete. Eighty-four million non-union workers often cite intimidation and 47 percent say that they would vote for a union in their company (*Business Week*, Sept. 13, 2004). Weekly wages of nonsupervisory workers have remained flat or declined in more than half of the sixty-five months since January 2001, according to the U.S. Labor Department.

Another huge player is the state of California, through its CALPERS and other public employees' pension funds. California State Treasurer Phil Angelides is one of many in other states who are concerned about corporate social performance. Phil launched the Green Wave Initiative—designed to bolster financial returns *and* create jobs *and* clean up the environment. In his imposing office in the state capitol in Sacramento, Phil explained his philosophy. "My responsibility as a trustee of our pension funds is simple, to get good returns, solid long-term returns so we can pay our pensions. Now I happen to be a big believer—as someone who was in the private sector for fifteen years before I was elected treasurer—that you can do well financially *and* do well by your society. So we have made a major investment in environmental technology and renewable energy. We call it our Green Wave Initiative, because we believe we can get a good return, and *also* we can foster clean energy, create new jobs, and make America energy independent. It is a way in which we can get our returns, but also do more good for our overall economy and society. I start with this premise: you can only be a successful investor if you're operating in a successful long-term economy and society."

Phil Angelides
CALPERS

Phil Angelides talks about the backlash from traditional Wall Streeters who still believe with conservative economists like Milton Friedman that their only concern should be growing the financial bottom line for investors. "The reactionaries want to label it social investment, rather than just looking at the facts. I think that we should not dismiss good investments because they also happen to be good for our society! The Green Wave, got the same kind of reactionary thinking—the labeling before the analysis of the facts. The facts show that global warming is a big economic issue. The facts also show that renewable energy and environmental technology is where America has to go in the twenty-first century. So the risk is always the risk of ignorance and reaction. But the way you overcome that is you keep pushing forward, you do your homework, and you prove the case. I think what I am proudest of as treasurer is that we've approached everything we have done in a businesslike manner." CALPERS pensions are a lot more secure than those of even profitable corporations, including Exxon Mobil's, which is underfunded by $11.2 billion and Lockheed Martin's by $4.5 billion (*Business Week*, May 29, 2006).

Ariane van Buren is senior manager of Investor Outreach for CERES (chapter 8). Ariane described how CERES was founded in 1989 to use asset allocation—namely investors and their money—as a way of influencing corporate behavior for the benefit of the earth and its inhabitants. "It's a coalition of some eighty-plus organizations, pension funds, investment organizations, labor pension funds, and environmental organizations—so a wide range of public interest organizations and investors (www.ceres.org). They use the status of those investors to try and leverage corporate behavior. Sustainability means the long-term economic viability of corporations in this country and abroad based on their use of the assets of the world. The Earth and its resources, not only the shareholders' money, and how companies deploy those resources, affects all of us. A number of pension funds are leading members of CERES, such as the state of Connecticut, the state of New York, the city of New York. Others of the California state government have led in the CERES coalition for many years. They have started an initiative through the treasurers of the different states because those treasurers are trustees of the public pension funds, which add up to trillions of dollars. These treasurers are getting right to the core of the problem and of the power structure.

They are the ones who call the shots. They then tell the funds what to do and then the funds tell the money managers on Wall Street."

Phil Angelides is deeply concerned about the recent corporate crime wave. "Over the last several years, America endured the worst wave of corporate scandal since the Great Depression. Our two pension funds lost $850 million on the WorldCom fraud alone. We have a responsibility to the pensioners we represent and to the larger economy to do everything we can to clean up the corporate boardrooms in our financial markets. That's why we have set tough new standards for investment banks and mutual funds that do business with us. It's why we are trying to get executive pay back to the realm of reality, where we reward true performance, not CEOs making $10, $20, $30 million a year, while they are looting their companies. That's our obligation as owners and we're remiss if we don't exercise our responsibility and our obligation." Phil continues.

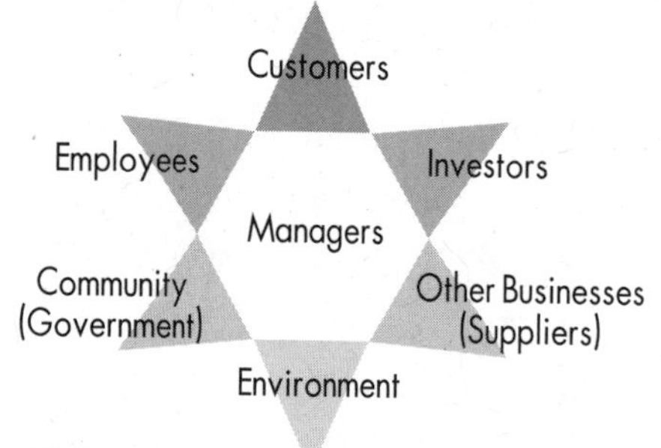

"In the area of corporate reform, I think there's no question that there's a whole body of leadership that understands that trusted markets in which millions of Americans are willing to put their money at risk are the best kinds of markets for their companies and the economy. There are some who still want to cling to the status quo. But the fact is, we can't go back to pre-Enron, pre-WorldCom times, it's not sustainable. We find some people now digging in, but I think the vast majority of progressive business people are willing to explore the frontiers of both financial and social responsibility."

In a global economy such as ours, this expanded stakeholder model affects companies in the United States, Europe, and elsewhere. Managers are taking account of this wider group of interests as part of their company's social contract with society. For example, a consortium of human rights groups including Social Accountability

International, led by Alice Tepper-Marlin, our stakeholder analyst, and companies including Nike, Gap, and Patagonia have devised a single set of global labor standards and a factory inspection system. Together with the Washington-based Fair Labor Association and London-based Ethical Trading Initiative, this consortium provides private-sector global standards analogous to those of the U.N. agency and the ILO (International Labor Organization) (*Business Week*, May 23, 2005).

Consumers often send messages in favor of companies practicing global corporate responsibility—the power is in their pockets. As Alisa Gravitz tells us, each and every purchase we make is an investment in our futures. "We can vote with our dollars to purchase and support companies that have products and services and good practices, or we can choose to withhold our dollars which is traditionally called a boycott. You can do that in an organized fashion where a national organization may call a boycott or you can simply refuse to purchase things. I always say vote with your economic choices, vote with your dollars, *and* vote with your voice, which means send an email or call the 800 number of a company. You can say, 'You know what? I'm not going to purchase from you anymore until you improve this environmental or this labor practice or this social justice practice.' And companies listen! If as little as 2 percent of their customers say the same thing about a concern they have they sit right up and listen because they cannot afford to lose you as a customer."

Company annual meetings used to be brief, quiet affairs, and asset managers routinely returned their proxies with votes for management and its slate of directors. The shift in shareholder expectations with regard to social responsibility has been gathering momentum since the 1970s. The early concerns for peace in Vietnam and social justice in South Africa surprised corporations and asset managers. Through the 1980s and 1990s, shareholders' concerns widened to workplace safety, diversity, living wages, human rights, environmental protection, and climate change. This new "politics by other means" was decried by traditional management, business schools, and Wall Streeters. Textbooks still said that business should stick to business—whose only social responsibility was to maximize shareholders' financial returns. A few diehards, including *The Economist* in its January 2005 survey, "The Good Corporation," stuck to this view, lamenting that wider definitions of corporate social responsibility are winning the day and claiming such efforts took money from shareholders. They ignore the fact that socially responsible companies are outperforming their conventional counterparts, and that shareholders also have goals other than financial rewards. Companies, 401Ks, and pension plans ought to reflect the views of their owners. Shareholders also have a right to organize, whether as pension beneficiaries, labor unions, or environmentalists—to assert their values in the marketplace—just as mainstream churches have done for decades. Now China's investors and consumers are active, too. Corporate social responsibility issues are covered weekly at ChinaCSR.com and www.EthicalMarkets.com. Chinese consumers organize on cell phones into team-buying by hundreds of shoppers demanding and getting huge price reductions on TVs, furniture, and other items (*The Economist*, July 1, 2006).

Most companies accept these new responsibilities as enlightened self-interest. They know the benefits they reap in enhancing their brands, in saving energy, materials, and money. Companies know that charters limiting their liabilities are privileged contracts with society. They know, too, that many groups are lobbying for reform of these charters to make companies more accountable (*ODE*, Jan.–Feb. 2005). While companies are not democracies, and shareholders' proxies are not the same as votes—the writing is on the wall. Today's demand on the part of shareholders everywhere for more democratically structured, accountable corporations mirrors the spread of democracies worldwide. Just as they have in the past, markets are evolving to meet new human needs and goals in this new century.

ROUNDTABLE
WALKING THE TALK

Jennifer Barsky, Simran Sethi, and Roberta Karp

Simran hosted Roberta Karp, senior vice president of Corporate Affairs and General Counsel for Liz Claiborne, and Jennifer Barsky, our stakeholder analyst and senior advisor of SustainAbility to discuss all these issues.

Simran Sethi: The International Association of Certified Chartered Accountants estimates over two thousand companies worldwide have produced separate annual reports on social, environmental, and ethical issues. Ethical Markets is committed to delving into those reports and finding out which companies are really walking their talk. Clothing retailer Liz Claiborne was lauded in *Fortune* magazine's list of most admired companies in March 2004. So, Roberta, can you tell me what practices make Liz Claiborne such a leader in the industry?

Roberta Karp: Liz Claiborne was working around the globe and realized that we needed to learn more about what working conditions are like. We always thought we went to the best factories in regions, but we learned that you had to really delve a little deeper and work with local NGOs and governments. We started to do that, and we took some risks by doing that.

Jennifer Barsky: It must be really challenging to monitor the factories. How do you do that, and do you have any partnerships that enable you to do that better?

Roberta: We do our own monitoring. What makes it really work is we are members of the Fair Labor Association. Independent monitors come into factories selected by the Fair Labor Association and then you essentially get a public report card by having your reports and summaries listed on the Web site.

Simran: How frequently are those factories inspected?

Roberta: We have a program once a year to inspect about half of them (internally it's more frequent). Then the Fair Labor Association will do a risk analysis of what factories they want to go in, so it's really within their control. So it's not every factory—we work in hundreds of factories—and there are other companies that are part of this and their factories are also monitored.

Jennifer: One of the companies that's come out with a really strong report in 2004 is the Gap. For the first time you have a company who's talking about the challenges that they face and basically saying, "OK, we're not perfect and actually, there's probably a lot of things still we don't know about and we really are having trouble controlling." It's set a new standard of disclosure for the industry. Is that something that Liz is planning on following suit on?

Roberta: I give the Gap tremendous credit for doing that because they didn't have to. It's very similar to the Fair Labor Association because these are the things that we're reporting on and on which others are looking at and reporting about us.

Jennifer: There's another thing that happened toward the end of 2004 and really started in 2005 and that is the end of the textile quotas in the United States and Europe. Obviously that had a pretty significant impact on a number of the countries like Sri Lanka or Mongolia who had these nascent apparel industries and are now competing with

China, Pakistan, and India. What have you been doing to help nurture some of the relationships that you've had with these factories in the countries that might be most impacted?

Roberta: For us, as a fashion company, we're looking for factories that can produce great product, that have a unique skill set. So we've been focusing more on the factory level. For at least five years, we've also been meeting with government officials, ministers of economy, asking what they are going to do to help their industries? Because it's not just about us doing something about it, it's really the whole community.

Simran: One of the countries to benefit so much from these WTO rules is China. They are very competitive in the low wages that they pay and they have a tendency to pull wages down overall. How do you face this challenge and how does it impact the other countries in which you work?

Roberta: Well, it is an issue for basic products where you're just going to compete on price. Wages are rising, but we work around the world to get different products.

Simran: Do you have any wage standards in place that say, "We ensure that we'll pay a living wage or a fair wage"?

Roberta: We monitor to make sure that workers are getting what they're entitled to get, that wages are being paid fairly, and that overtime is being paid. Just monitoring has been a challenge. You'll see that in some of the reports, some vendors have two sets of books. It's our job to make sure it's not happening and they're paying their workers. We've had tremendous buy-in and support from our vendors. They know we're serious about it. It helps to have more companies go out there so we're not a lone voice. We're certainly a strong voice in the factories that we use because we share them with other vendors as well.

Jennifer: It sounds like you're constantly facing the challenge of balancing shareholder needs, the profit motive, with corporate social responsibility efforts that you're undertaking.

Roberta: Corporate social responsibility is really part and parcel to the overall business practice. We're a corporation, we have a reputation and so we have to deliver good products and we have to do it in a responsible manner. You don't want surprises, you have to protect your assets and so it's one of the many factors that are evaluated. To us it all just works together and makes sense.

Simran: What kind of code of conduct do you have in place?

Roberta: Our first one was in 1994—maybe even in 1993—and I wrote it with the help of colleagues and it's evolved. The Fair Labor Association has a standard code of conduct. The codes are just part of the story. Monitoring against them, the discussions, the meeting with the workers, having secure channels of communications to understand what's happening is key. Five years from now, I think we'll be even further along, and I look forward to being part of seeing the world move in that direction.

TEN

Transformation of Work

Over the last three decades, the amount of time Americans spend at their jobs has risen steadily. According to the U.S. Bureau of Labor Statistics, the average American worked 1,801 hours in 2002, five-and a half weeks more than in 1976. Add to this, the fact that the average American gets just over ten vacation days a year. And with BlackBerries, cell phones, and laptops, work is only a click away.

While these communications technologies hold the promise of freeing us from offices and saving time, *Business Week* economist Michael Mandel points to the need for corporations to restructure their hierarchical management to create this freedom—especially for their knowledge workers, in his cover story "The Real Reasons You're Working so Hard—and What You Can Do About It" (Oct. 3, 2005). All this is symptomatic of dysfunctional corporate organization still locked in the industrial age. Verna Allee (chapter 1), author of *The Knowledge Evolution*, points out in today's information age productivity and innovation are often the result of value-creating networks of employees within companies. All this puts a premium on employees collaborating across all those silos and departments still the norm in old-style corporate hierarchies—meaning extra time on those e-mails and BlackBerries to overcome. So time pressures have increased such that, according to a recent survey by McKinsey of more than seventy-eight hundred managers around the world, 25 percent of managers at large companies say their e-mail, voice mail, and meetings are nearly or completely unmanageable. Worse, nearly 40 percent of respondents reported that they spend a half to a full day a week on communications that are not valuable (*Business Week*, Oct. 3, 2005). The Career Innovation Company of Oxford did a survey in 2006 of the United Kingdom's global knowledge workers. It found that 55 percent were dissatisfied with their sense of achievement at work; 54 percent were dissatisfied with how their skills are being used; and 56 percent were dissatisfied with their ability to achieve a work/life balance. U.S.-based expert on work/life balance, Nella Barkley says that large firms are beginning to understand, "but only slowly" (*The Economist*, June 17, 2006).

Many Americans may feel that they can't afford to take their vacations given the increase in job outsourcing and the fact that wages for nonsupervisory jobs have barely kept up with inflation. While managerial salaries have risen some 30 percent since the 1980s, wages for workers have remained flat. A recent experiment to improve GDP national accounts to better measure the contribution of intangibles, like research and development, by counting such outlays as capital investment leaves little doubt that workers in the United States are being left behind and the share of income going to corporate profits has increased. Such changes would also show that, according to a recent study by Federal Reserve Board economists Carol A. Corrado, Daniel E. Sichel, and University of Maryland economist Charles R. Hulten, "The hidden earnings from these knowledge investments have not been shared equally with workers" (*New York Times*, Apr. 9, 2006). Technology and automation has made many tasks obsolete and our U.S. tax code encourages this drift by making capital equipment relatively cheap compared with human workers. Outsourcing, like cheap Wal-Mart goods, benefits consumers. But trade deals like

NAFTA, between the United States, Canada, and Mexico, which promised more U.S. jobs, has been responsible for the loss of five hundred thousand jobs in the United States between 1994 and 2002, according to the U.S. Labor Department.

Jeremy Rifkin, president of the Foundation on Economic Trends, wastes no time in launching into these complex issues. "Let's cut right to the bottom line. We have all these sophisticated technologies, information technologies, intelligence technologies, and they're increasingly replacing human labor, in every field, in every industry, in every sector. When I wrote *The End of Work* there were eight hundred million people underemployed and unemployed in the world. That was ten years ago. Now, there are a billion people underemployed and unemployed in the world. Unemployment is a global problem. It's also a global opportunity. We are experiencing one of those great thresholds, one of those changes in history where we're going to have to redefine work. The cheapest workers in the world, in factories, offices, in professional suites, are not going to be as cheap as the intelligent technology that can replace them. Forty years ago when I was a student at the Wharton School of the University of Pennsylvania, where I now teach, a third of the American labor force were blue-collar workers. Today, less than 17 percent of Americans are in factory jobs. Yet in the United States we still number one in manufacturing. Thirty years from now, factory jobs will be virtually eliminated across the world. Even China has eliminated 15 percent of all of its factory workers in seven years. Why? The cheapest Chinese worker, and they're pretty cheap, is not as cheap or as efficient as intelligent technology that's automating those factories across China.

Jeremy Rifkin
Foundation on Economic Trends

"The industrial age has been characterized by mass human labor working side by side with machine technology. Machine technology is still substituting for that labor and we're moving to boutique workforces. It is true we're developing new goods, new services, new skill levels, and there's going to be new opportunities. But there will not be mass labor again. It is a tremendous change for the human race, and we're not prepared for it."

I also covered this issue in all my books, as it unfolded over the past thirty years. E. F. Schumacher predicted job loss due to automation in his best-seller, *Small Is Beautiful* (1973), which pointed out, as I did, that industrialization was about labor-*saving*—and would eventually be unable to employ enough people to be a viable model of development in poorer countries. A student of Mahatma Gandhi, Schumacher said, "We need mass-production less than we need production by the masses."

The Industrial Revolution saved labor by raising per capita productivity through investing ever more capital in technology and innovation. Schumacher correctly predicted the looming global unemployment problem in the United States: first farming was mechanized and former farm hands went into factories and the cities. Then factories became progressively automated and their workers migrated into white-collar jobs in the growing services sectors of the United States and other maturing industrial societies. Today, services are being automated by banks, supermarkets, phone companies, and their endless telephone trees—or outsourced to India and other lower-wage countries. Northern European countries outsource such call-center jobs to Spain—a lower-wage member of the European Union. Japan outsources to a largely Japanese-speaking city in northern China. Yet social norms, economic textbooks, and government policies have not kept up with these changes. Patricia Kelso, who with her late husband Louis O. Kelso invented Employee Stock Ownership Plans (ESOPs), is working to increase opportunities for employees to share in company profits. This kind of social innovation generally follows technological innovation—usually with considerable time lags. Visionaries like the Kelsos often find their proposals and ideas rebuffed as too unfamiliar, or because they challenge old structures and special interests. I have been a supporter of the Kelsos proposals since the 1970s and can attest to their widespread rejection—not only by companies and government agencies, but even by labor unions, whose economists were trained in the same schools as those working for businesses and government.

Patricia Kelso is now president of the Kelso Institute in San Francisco, which promotes employee stock ownership, economic democracy, and new models of twenty-first century capitalism in Eastern Europe, Russia, and China. Patricia explains, "The basic question in economics is, how do people earn a living? Conventional economists reply: through labor, work, and wages. Louis Kelso's answer is: through work to the degree that the economy *needs* your labor but through capital ownership to the degree that it does *not*. Look at an oil-drilling platform out on the Gulf of Mexico or Europe's North Sea. You will see a towering, costly capital instrument and a *few* people. They are working for wages but that capital is working for its *owners*. Now the question is, where do customers get the money to buy? They can only get it through their *labor* or through their *capital*—or as Kelso would advise, through *both*. The bottom line is working people need to augment their labor income with capital income. We run our economy on the myth that there is only one factor of production, labor. In reality, we produce wealth through two factors: labor and capital. We still try to distribute the earnings of two factors through one factor, namely labor, and it just won't work. Louis Kelso began to think about how capital was financed. It was financed from savings.

Patricia Kelso and Louis O. Kelso
Kelso Institute

In reality it's not financed from savings. It's financed from *credit*. But this credit is only given to people who *have* savings."

In Louis and Patricia Kelso's *Two Factor Theory: How to Turn 80 Million Workers into Capitalists on Borrowed Money* (1967), they foresaw how the continual drive for labor productivity (i.e., more output per worker) would lead to more automation *and* unemployment. How would purchasing power be recirculated so unemployed people could buy all those goods pouring off the production lines? Industrial societies have hardly faced up to this problem—let alone solve it. Kelso predicted that it would continue to be covered up by governments becoming employers of last resort, which would lead to ever more warfare, workfare, and welfare—rather than making sure that every worker was enabled to become a capitalist.

Rich Ferlauto (chapter 2) has thought deeply about how these issues affect the members of his union, AFSCME. "When the Enron scandal broke, our president understood that something needed to be done directly to protect the retirement assets of AFSCME members. He decided to dedicate some significant resources to organizing the assets of AFSCME members. The Enron scandal and the two years following the scandal caused public retirement systems, exclusively, to lose $300 billion in assets. That translates to about 15 percent of their total assets. It means that the public retirement systems, as in people's individual 401Ks and mutual funds, took a huge hit that they still haven't recovered from." The United Steelworkers now teams up with green groups like the Sierra Club to promote "Good Jobs and a Clean Environment."

Today, it seems harder than ever for workers to increase their share of the productivity pie to save, or secure retirement benefits. Companies in troubled sectors, including airlines, steel, telecoms, and automotive companies, have gone bankrupt. Their pension plans were turned over to the federal Pension Benefit Guaranty Corporation, which has taken over payments to beneficiaries. Workers have lost significant percentages of those promised benefits. Meanwhile, hundreds of companies shifted from defined-benefit pension plans to defined-contribution plans funded largely by employees through 401Ks, IRAs, and the like. The Kelso's ESOPs allow workers to acquire stock ownership in their companies *out of the capital earnings of those companies*—rather than having to save up to buy those shares. Patricia explains, "We had no mechanism whatsoever until Louis Kelso invented the ESOP which gives access to capital credit to people who have no savings. The ESOP is that mechanism. It enables the employees of successful corporations to become major stockholders in that corporation and pay for their capital out of the earnings produced out of that capital." This revolutionary approach to expanding capital ownership took many years of patient effort by the Kelsos and their supporters, until ESOPs were enacted in the 1980s. Now, employee-owned companies have proved themselves, as documented in *Equity* by C. Rosen, J. Case, and M. Staubus (2005).

Bernie Glassman succeeded in integrating his spiritual beliefs into an effort to expand the pie for workers. The Brooklyn-born native is a high-ranking Zen priest and the founder of Greyston Bakery, a Yonkers company empowering the community one cookie at a time. Bernie talks about his mission, "At the time that we started Greyston, the biggest problem in Yonkers was homelessness. Yonkers had the highest per capita homeless rate in America. Yet, that was part of Westchester County, which is one of the wealthiest counties in the country. So I said, can we put an end to homelessness in Yonkers? The obvious thing about homeless folks, is that they need housing! For the ten years prior to the time that I had started to do my work, Westchester County had continuously changed zoning rules to make the size of a lot larger and larger. This meant there was less and less housing. In that ten years prior to my

Bernie Glassman
Greyston Bakery

starting the work, there were zero new apartments of any type built in Westchester County. So the housing stock went down. That was also a time that was feeling the effects of the Reagan administration cutting budgets for affordable, federal housing. So multiple things were in effect to make homelessness rise very rapidly in Yonkers.

"What went into the philosophy of starting the Greyston Bakery? It was to provide jobs for the folks who we were working with. Not only the homeless. Many of the initial people that we hired were not homeless but had been unemployed for long, long periods of time. Perhaps some of them were dealing drugs, and had given that up out of fear of being killed, or for other reasons. The majority of folks that we brought into the Greyston Bakery were people who had not had any skills in baking or in maintaining a job. So the Bakery was to provide jobs, and it was also a place to enable folks, who were making money now from their jobs, to start looking beyond themselves, beyond their families, and into the society to have a bigger goal. The Bakery was to make money. So, the bottom line from the beginning was to make a profit, and to serve the community. Both were a part of our bottom line."

> The Greyston Bakery was to provide jobs, it was also a place to enable folks, who were making money now from their jobs, to start looking beyond themselves, beyond their families, and into the society to envision larger goals.

Wendy Powell was an administrator at Greyston Bakery and shares her perspective. "It was ten years that I was employed at Greyston—as well as ten years since I first obtained permanent housing. I was recently graduated from earning an associates degree. I had a four-year-old son and we were homeless. At Greyston, they have a pretty specific process to get housing. You need to go through a process to make sure you're serious about wanting to make some changes to your life. So I did it and it worked! I went through a series of really nice promotions while I was at Greyston's daycare center. I feel really proud to say during their times of transition I was able to step up and keep everything grounded. And I was only able to do that because the company trusted me. They gave me the responsibility because they had faith that I could do it. Very honestly then I didn't believe I could do it. At Greyston, they are able to extend opportunities to people who might not normally get this kind of opportunity. It's not a huge high paying job like some people just coming in might like it to be. But it's definitely an opportunity to show up somewhere each day and feel respected and to earn an honest wage."

The United States, Canada, Japan, and mature industrial countries in Europe have transformed themselves over the past century into predominately service-based economies. Such economies become information-based and knowledge-intensive. Thus, knowledge has become an important new form of capital. Few economic textbooks and accounting protocols have devised ways to value information and knowledge correctly. Information and knowledge do not conform to the old economic textbook models of scarcity. Economists assume that fundamental competition for scarce material resources drives economic growth. This economic view still makes sense on the material plane: oil will run out in this new century, fish stocks are endangered in all the world's oceans, minerals in more remote areas get more expensive to mine, and thus the price for these commodities rises. But information is not scarce. If you give me information, I am enriched—and you still have it, too! As industrial economies matured into information and services, the necessary revolution in economics didn't happen. While we await this overhaul from the economics profession, other approaches, such as those of the Kelsos, Dana and Dennis Meadows, Amory and Hunter Lovins, Fritjof Capra, Elisabet Sahtouris, Riane Eisler, myself, and many other systems analysts have filled the void. In the Calvert-Henderson Quality of Life Indicators, I cover "The Politics of Productivity Measurement" and still call for the needed changes in economic

Mature industrial countries have transformed themselves over the past century into predominately service-based economies. Such economies become information-based and knowledge-intensive. Thus, knowledge has become an important new form of capital.

models, including reclassifying education from consumption to investment (click on "Current Issues" at www.Calvert-Henderson.com). *Business Week* economist Michael Mandel reached the same conclusion in his cover story "Unmasking the Economy" (Feb. 13, 2006).

In *The Future of Knowledge; Increasing Prosperity Through Value Networks* (2003), Verna Allee also focused on this great industrial transition to knowledge-based economies. She relates how huge social changes have been driven by innovative companies—and also changed the game for many other firms. "This transition is what a lot of companies have missed. They are still trying to get their competitive advantage based on technology, based on cost-cutting, based on efficiencies, based on the production line, on value chain. They are missing the phenomenal amount of social innovation that is going around knowledge networks, practice communities, the rise of the civil society organizations—all finding each other and networking across old boundaries. These companies are completely ignoring the intangible aspects of the business model, and most of them are really struggling. If you want to create a business that endures over time and particularly if your business depends on innovation and fresh ideas, then you need to have people coming to you who trust you with their good ideas.

"We did a study in 1997 using Value Network Analysis to look at some of the emerging e-commerce businesses. We saw the way that they were using knowledge exchanges very consciously and deliberately with all of their stakeholders in their business groups. They were utilizing their customers' knowledge, suppliers' knowledge, sharing their own strategic goals with each other, with their partners. Three companies that understood the new terrain were Cisco, eBay, and Amazon. Amazon at that time was starting to get a little bit of business press, but back then no one was talking about Cisco or eBay. Yet, we were absolutely blown away by the number of intangible exchanges they were leveraging in their business models. What's really interesting is that all three of those companies survived the dot-com crash just fine."

We are all learning that information-rich, networked societies and their markets run on *trust*, another crucially valuable factor that economists ignore in their models. Furthermore, the new knowledge factor of production exists in the heads of employees—not in some safe at company headquarters. This is why so many innovative companies strive to keep their employees' trust and keep them happy and productive in so many new ways, from ESOPs to daycare, gyms, sports facilities, and all manner of other perks.

Gary Erickson, CEO of Clif Bar manufacturers of organic snacks for athletes, in California explains his policy. "People have noticed that we are doing business differently, that taking care of our people is one of our bottom lines, and taking care of our community, and giving back. We allow our employees to do community service during work hours, and we shoot for a goal of 2080 hours a year, which is the equivalent of one full time person's job. We donate a lot of Clif Bars to different food banks. We believe that the workday can be more than just to come in here and grind out the work. So we have a world-class gym, and adjoining that we have this beautiful dance floor where we hold yoga classes, karate, kickboxing, and hip-hop. A lot of companies around have gyms, but we have such a high participation rate, we know it is working. And our trainers are the kind that encourage people. They will go out and persuade workers to take a break. People need to recover during the day. That's what athletes do, you stress your body, then you recover. If you work straight through eight hours a day you are going to burn out.

Gary Erickson
Clif Bar Company

"We do health tests here. We have a concierge service, where you can wash your car, do your laundry, have your hair cut on Thursdays, your wet cleaning (which is *environmental* dry cleaning). We have trips where we go rafting, climbing, or skiing. We have fun parties in our beautiful auditorium that holds up to three hundred people. Every week we meet in the auditorium as a company, with a lot of open communication. Because we are a private company, we don't offer any stock options, so the way we do that is have an annual incentive plan where we offer a profit sharing program."

Gary certainly understands how to run a company and prosper in an information-rich, networked society. Clif Bars compete successfully against giant food conglomerates. Eighty-five percent of Americans believe that living in a fair and just society better describes their concept of the American dream and growing U.S. inequality was the subject of a special report in *The Economist,* June 17, 2006 issue. People want to make a difference in the world and are looking for work that provides a living wage as well as something more meaningful.

Sociologist Paul Ray (chapter 7) tells us more about those over fifty million Americans of diverse backgrounds that he and coauthor Sherry Anderson call the Cultural Creatives. They come in all ages and incomes and are best defined by their *shared values*. Paul explains further, "The Cultural Creatives are a part of the population that the news media don't seem to want to acknowledge. So you don't see their faces on TV, you don't hear about them in the newspapers, you don't hear their values being talked about in the workplace. One of the big oil companies, for example, for whom I did some consulting on a public affairs campaign, took the environmental questionnaire that I used to set up the focus groups. They gave it to their employees, and found to their shock, that over 70 percent of their employees completely *agreed* with the Culture Creatives on environmental issues! They've slowly started to change because they figure if their employees don't even agree with management, maybe there is something important that they need to pay attention to."

This fifty million strong cohort of the U.S. population is beginning to change the stale politics of both Republicans and Democrats. New alignments around their deeper knowledge and experiences of living in information-based networked societies are swelling the ranks of independent voters, now some 40 percent, who deeply challenge obsolete industrial economic assumptions. Republicans and Democrats are now both minority parties with less than 30 percent of voters, so they are soul-searching about how to update their respective messages to embrace the realities of most U.S. voters' lives. Canadian pollster Michael Adams describes these realities in his *American Backlash* (2005).

Lynne Twist, author of *The Soul of Money* (2004), is another new opinion leader. "You know we have this kind of running dialogue in our heads that *there's not enough*. There's not enough time, there's not enough money, there's not enough love, there's not enough sleep, there's not enough this, there's not enough that. This kind of mentality justifies horrendous behaviors in society. This consumer culture is really crushing our capacity to know who we are and to really live within any kind of boundaries. What I recommend is what I call the surprising, stunning, and shocking truth of *sufficiency* or *enough*! And in a global society that's now become a consumer-monetized society everyone is targeted, studied, treated, and talked to as a consumer, and nothing more than that. There is a hunger for more centered ways of understanding our relationship with money. There's a desperate cry—that's drowned out by advertising—to get out of this consumer culture!"

Lynne Twist
The Soul of Money Institute

Lynne is describing exactly the same feelings revealed in the many surveys conducted by Betsy Taylor's Center for A New American Dream (chapter 1). My view is that these new lifestyle concerns for personal growth, and environmental awareness are creating that Attention Economy where time is valued more than money, which I described in *Beyond Globalization* (1999) and *Planetary Citizenship* (2004).

Workplaces have changed. No more lifetime jobs with benefits or loyalties tied to one company. Americans have seen the part-timing, downsizing, and temporary revolutions. They now face the outsourcing revolution, where manufacturing jobs migrate to Mexico, China, and other low-wage countries. The International Confederation of Free Trade Unions (ICFTU) estimated in March 2006 that more than twenty-five million private civilian workers and 6.9 million federal, state, and local employees in the United States are excluded from the Labor Relations Act. For those workers who do enjoy the right to organize, there is insufficient legal protection against anti-union discrimination. ICFTU Secretary General Guy Ryder, noted in the report, "The credibility of the United States, which takes a strong international stand on human rights issues, is severely damaged by the lack of protection for working people, especially within its own borders. This only encourages other governments to seek a competitive advantage in global markets by violating the fundamental 'rights of their own workers' (InterPress Service, Mar. 15, 2006). Now U.S. high-tech and professional jobs are leaving the country—even R&D is going offshore where Indian and Chinese PhDs earn less than one tenth of their American counterparts.

Yet, Americans are a flexible and entrepreneurial people. Some are willing to commute three hours a day (*Business Week*, Feb. 21, 2005). They job hunt, upgrade their skills, and change careers. Some fourteen million U.S. workers telecommuted at least part-time in 2004, and another seven million run businesses from home according to the U.S. Labor Department. Others have started businesses on the Internet and new businesses make the United States still the most dynamic economy in the world. The twenty fastest growing professions include high-level services such as environmental engineering, network and data analysts, personal financial advisors, software engineers, medical professionals, counselors, and social workers (*Fortune*, Mar. 21, 2005). Thousands of entrepreneurs in the United States and Europe run businesses on eBay (*Business Week*, Apr. 3, 2006). The homeshore trend, an alternative to offshore hiring, grew 20 percent in 2005 to 112,000 jobs and will hit 330,000 jobs by 2010. Most "homeshorers" are educated, stay-at-home moms living in rural areas, employed by virtual call-center providers like Alpine, Access, LiveOps, Willow, and Working Solutions (*Business Week*, Jan. 23, 2006).

Yet, the price we pay for all this dynamism can fall on families and communities in the loss of time and security. To recap the long history, industrialism released innovation and technological change with its goal of saving labor by using machinery. The resulting efficiency gains first mechanized agriculture. Displaced farm workers went into factories. Then automated production lines drove workers into low-paying service jobs. (There are twenty-eight million working poor in the United States earning $18,800 per year—the poverty line—mostly in service jobs.) Tax codes still make machines cheap and humans expensive. To remedy this problem, many groups are trying to shift taxes from payrolls and incomes to pollution, waste, and resource-depletion. Europe leads in such green tax shifts. In the United States, conventional industries offer entrenched opposition to such a logical shift of taxes—for example oil companies would no longer receive oil depletion allowances but would be taxed to account for oil's pollution. Bill Drayton (chapters 2 and 5) founded the Washington-based NGO Get America Working, which calls for green tax shifting to save resources and provide millions of new jobs (www.getamericaworking.org).

The great, only partly fulfilled promise of the Industrial Revolution was leisure societies—with ever-shorter work weeks. People would enjoy the arts, sports, self-improvement, learning new skills, more travel, and vacations—and this would create a whole new economy. In the 1970s, the debate was about how to maintain purchasing power to buy the floods of goods and keep those automated production lines humming. How would laid-off employees have incomes?

Three new ideas proposed to meet the challenge: 1) a guaranteed minimum income for all, a proposed negative income tax; 2) guaranteed jobs; 3) the Kelso's idea: employee stock ownership plans. Many agreed that if the machine takes your job, you'd better own a piece of that machine. What happened? Guaranteed jobs became law in India. The negative income tax guaranteed income was a nonstarter ("no workee—no eatee"). But ESOPs are flourishing with eleven thousand companies now owned by their employees like Chroma Technology in Vermont. So an ownership society is possible, and we are slowly witnessing a transformation of the workplace that fits an enlightened twenty-first century model.

ROUNDTABLE
WALKING THE TALK

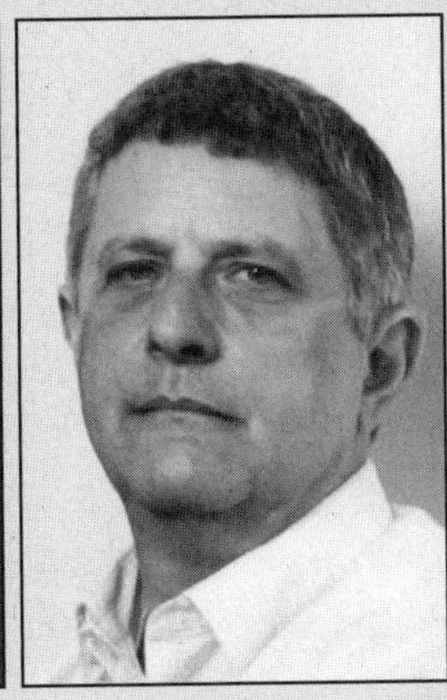

Simran Sethi, Paul Millman, and Paul Freundlich

Simran hosted Paul Millman, CEO of Chroma Technology Corporation, and our stakeholder analyst, Paul Freundlich, president of the Fair Trade Foundation, a founder of Co-op America, and board member of the CERES Coalition, to look at Chroma Technology, an ESOP company.

Paul Millman: Chroma created an ESOP because it was very clear to me that the only way I could function in that company is if all the employees owned the company.

Paul Freundlich: Do you think it motivates you as a company to be more effective, more creative, and more innovative?

Millman: Yes, everyone who works there is responsible for their job and doing it well because we all own it and because we're going to get rewarded by our work.

Simran Sethi: You've chosen to stay in Vermont instead of moving to New Hampshire, which was an opportunity that was recently afforded you. Why is that?

Millman: Partly because of Vermont values. In addition to Vermont Businesses for Social Responsibility, Vermont also has the Vermont Employee Ownership Center. Vermont and Ohio are the two states that recognize employee ownership as viable business organizations.

Freundlich: Chroma isn't a public company. Do you think your company works better because it's employee-owned rather than if you were publicly traded and had investors?

Millman: In the vernacular of the times, they're talking about the ownership society. Well, we're the real ownership society. We really do own ourselves. And when you own yourselves, everything that happens in your company, good or ill, affects you. This is very different than if you're an outside shareholder and what you want is the company to make money regardless of what kind of practices it might entail. We do a lot of prototype work for scientists at what amounts to a loss. It's really not a loss for us because we're paying ourselves anyway. Yet, we could conceivably make more money if we charged more for that work. But we're talking about scientists who don't have a lot of money to spend on tools for their trade. If we had an outside group of shareholders would they let us get away with doing it at a loss or doing it for less than if we charged full price? I don't think so.

Freundlich: You have, essentially, the luxury to look ahead and say we want a company that's going to be viable in the marketplace two years from now, five years from now, ten years from now. Because your jobs depend on it, this is very different than the CEOs of many of the publicly traded companies, which have to deal with the quarterly earnings statements and the pressure that engenders. [These pressures from Wall Street security analysts led to the downfall of many companies. They began to fudge their figures to try to show those analysts continuous rates of growth. Such creative accounting has now been made illegal under the Sarbanes-Oxley law and hundreds of companies have had to restate their earnings at lower, more realistic levels.]

Millman: It doesn't mean there aren't pressures on us, that there isn't a temptation out there to make the big killing at some point in our history.

But mostly what we want is to create a company that continues to exercise the values we have for as long as we possibly can.

Simran: What are you doing to reduce the environmental impacts of the manufacturing process?

Millman: We use a tremendous amount of water in our manufacturing process. In 2004, our new building was created with a system that recycles all of the water. So we've reduced the amount of water we use from hundreds of millions of gallons to something less than forty thousand gallons. We've also improved energy conservation and we got an award from Efficiency Vermont. We use a lot of energy for manufacturing high energy-using instruments. We try to do it as efficiently and in as environmentally friendly ways as is possible.

Employee-owners of Chroma Technology Corporation

ELEVEN

Clean Food

INDIVIDUALS AND ENTERPRISES FROM ALL OVER THE WORLD are transforming the way we grow and consume foods through organic, sustainable, delicious, and profitable methods. The explosive demand for nongenetically modified, organic, pesticide-free food reflects a new awareness of what we eat and how our food is raised—prompted in large part, by frightening stories that keep emerging in the media about the safety of the food we eat. Mad cow disease, mercury in fish, or dioxins in breast milk forcibly show that the industrialization of our food supplies, use of persistent pesticides, and energy-intensive, monoculture farming is unsustainable and unhealthy. In the United States, the organic market has grown 20 percent annually for the past five years, compared with only 3 to 4 percent for the industry as a whole and is projected to reach $30.7 billion by 2007. A 2004 study by the Food Marketing Institute found 56 percent of Americans strongly agree that eating well is a better way to prevent health problems than taking drugs. The U.S. government in 2005 issued new Dietary Guidelines, which encourage people to rely on a diet of fruits and vegetables (nine servings a day), whole grains and low fat dairy products, less sugar, salt, and trans fats—over the protests from various segments of the food industry (*U.S. News and World Report*, Jan. 24, 2005).

Rodale Press and Institute in Emmaus, Pennsylvania, have led the way for decades, with their popular magazines, *Organic Gardening* and *Prevention*. Rodale is still a family-owned business, chaired by Ardath Rodale. Her daughter, Maria Rodale, is a top executive, and son, Anthony Rodale, is chairman of Rodale Institute, the company's nonprofit research arm. Anthony describes the Rodale philosophy, "Our intention at the Rodale Institute is to have more farmers transition into organic farming to meet the demand. We're committed to helping to train and educate farmers to make the transition to a much more environmentally friendly, regenerative way of farming. In establishing the scientific understanding for organic farming, there was a realization that the soil is actually improving while we are using organic methods and techniques."

There are currently a little over twelve thousand organic farmers in the United States representing about 5 percent of the domestic farming population. This percentage can't keep up with the demand for organic food, so 10 percent of the organics that are consumed come from abroad. And this presents an interesting dilemma—people want to eat well but also want to support local farmers and reduce the amount of fossil fuel it takes to transport food from the farm to their forks. Anthony Rodale explains further, "Food, when it's harvested, begins to change. The life is cut once the food is harvested from the ground. The nutrients change as well. If we eat foods that are grown locally, the nutrients are retained because the food hasn't traveled a long distance. The local and regional agriculture and food systems provide an incredible security for each region as well. So it's something that people can feel good about." Nutrition expert, Professor Marion Nestle of New York University agrees in her latest book *What to Eat* (2006).

Meanwhile, in today's globalized economy, food, such as endangered fish like Chilean seabass and out-of-season fruits, travels thousands of miles to reach consumers. Gary Hirshberg, president and CEO of Stonyfield Farms, now a division of French food giant Danone, sums up the dilemma, "If we all could purchase everything within fifteen miles, or even a hundred miles, and it was organic, meaning produced in a manner that's consistent with natural principles, then of course we should do that. Unfortunately, 85 percent of humanity is urban, and most people can't do that. So, it's important to us at this stage in societal evolution that we get organic foods out there. Ultimately as fossil fuels costs go up, as transportation miles become really expensive, as we start, like Europe has for decades, recognizing the true costs of transportation, I think agriculture will become more and more local or bioregional."

Thomas Fricke, cofounder of Forestrade, Inc., addresses the need for universal standards and is a pioneer in the organic food industry. "We have to be certified organic, that means we're required really to work at the grass roots and from the ground up. So we work with, now, over eight thousand individual farmers in hundreds of locations around the world. Organics is a whole systems approach. It really allows us to make a connection between producers and consumers. It enables people to really make a personal relationship and to really have an impact on protecting and preserving the way of life of the producers. This is helping to protect some very important natural eco-systems. So we really need to be attentive and very much engaged with local cultural practices, mores, and values."

Thomas Fricke
Forestrade, Inc.

Yet, there is still a widespread belief that small producers can never supply sufficient food in today's world—and that only industrial agriculture and food production can meet humanity's needs. Anthony

Rodale disagrees. "Many people believe that organic farming cannot get the appropriate yields. Well, we found through one of the worst droughts ever on our farm that the organic yields were superior to the conventional yields. Also, this year we realized—being a very wet year—that the organic crop actually outperformed the conventional crop. In general, an average year, we see that the crop yields are pretty much the same. Supermarkets have only been in existence for forty years. Before that, people got most of their food locally and regionally. They lived within seasons and cycles. Today, you can get foods at any time of the year because people are demanding it. The importance of the local and the regional movement that is happening today is people wanting to reconnect. A person buying a great cheese that's made locally and has a story behind the farm provides an incredible educational awareness. People love to talk about food, and it provides another great way to connect other people to foods that are grown locally, organically."

Witnessing the exponential growth of the natural food industry, Wal-Mart now sells organic food and is already the biggest seller of organic milk. Agribusiness has also entered the arena, creating its own healthy products and buying up smaller organic and clean-food companies. Seeds of Change is now owned by M&M Mars, Ben and Jerry's is owned by Unilever, Boca Foods is owned by Altria, the tobacco firm, formerly known as Phillip Morris. Horizon Organics and Silk Soy Milk are owned by Dean Foods. Gary Hirshberg explains how the sale of Stonyfield Farm to consumer products giant Danone has affected his business. "One of my proudest achievements in twenty-two years of running this company has been the marriage with Group Danone, and I use the word marriage intentionally. This has most definitely not been an acquisition. We have an independent Board of Directors. I still have a majority of board seats, even though I'm only now a 20 percent owner. But truly, in three years, Danone has proven that they are much more sitting in the bleachers cheering us on than anything else. They put no money into the company, the money that they used—that they spent—was to take our investors out, and they got an excellent return on investment. Now I spend the bulk of my time working on a number of organic initiatives globally. Danone is a global company. I really do believe that we are one of those little spores inside Danone that will really greatly change the way that they do business. I believed deeply from the beginning that by ourselves, Stoneyfield was not going to save the planet. That what we needed to do was to serve as a model, that we can show large companies, established companies, the way. Frankly, McDonalds, Coca-Cola, General Mills, Danone can do more good with one purchase order than I may do in a lifetime! And indeed, if any one of those companies made a real aggressive commitment to organic, it would change the world. Stoneyfield is changing the world also, just on a much slower scale. I really believe it's our moral obligation and mandate to change these companies. It's a great opportunity for these companies, and Danone has really recognized that. That's why they left us so independent."

Gary Hirshberg
Stonyfield Farms

Danone is a target of interest for acquisition by Pepsi Cola. Now that large companies are buying up organic producers, they have lobbied the Department of Agriculture (USDA) to lower its organic standards to allow certain synthetics in the preparation, processing, and packaging of organic foods. The Organic Consumers Association (OCA) fought hard against lobbyists from Kraft, Wal-Mart, Dean Foods, and other big firms in an effort to retain their hard-won standard, the Organic Food Production Act—OFPA. Despite the legitimate concerns of organic producers, lobbyists from Kraft, Dean, and other big companies who now control the Organic Trade Association won the fight. The lower organic standards came into force in mid-2006 (*Business Week*, Apr. 10, 2006). Over the protests of consumer groups, U.S. Agriculture Secretary Mike Johanns even refused to fill the vacant consumer slot on the National Organic Standards Board (www.cspinet.org/integrity).

Clif Bars (chapter 10) has succeeded without corporate backing and sticks to its high organic standards. The company's Luna Nutrition Bar For Women has outsold Nestlé's power-bars to become the best-selling energy and nutrition bars in grocery stores. This supports founder Gary Erickson's commitment to private ownership. "I got Clif Bar started because I was a consumer of a product that wasn't working for me. I was bicycle racing, and doing tours in the Alps, and one day I was on a long bike ride. I took along on this ride six of the only energy bar on the market in 1990. After about 125 miles on this ride, I had eaten five of the six bars, and I had another fifty miles to go and I needed fuel. I needed to eat something, but I looked at this last bar, and I just couldn't eat it. So for me that was both a moment of inspiration and an epiphany. I could make something better than this, and I did!"

Clif Bar grew from a $700,000 business in 1992 to $40 million by the year 1999, in annual sales without any investors. Gary adds, "We are privately held, with two partners with equal fifty-fifty ownership. Our two largest competitors, Power Bar and Balance Bar, were purchased by the two largest food companies in the world: Nestle and Kraft. We're this small Berkeley-based, California-style hippie company competing against the two largest food companies in the world! The experts from the outside world were telling us, we were never going to be able to compete. They said, you did a great job, you're an entrepreneur, you built your company, but it is time you think about exiting or bringing on some investors. So in an uncharacteristic moment for myself—because I am willing to take on risk—I fell into the trap of believing that we couldn't make it. The company was put up for sale. Here I am, a guy who used to live in a garage, down the street here in Berkeley, with no heat and no bathroom, making $10,000 a year. Now I have a chance to walk away with $60 million. At the last minute, literally two hours before walking away with the money, I took a walk down around the block. Half way around the block, in my heart, I decided not to do it. Today, I feel the only way to make sure that this vision stays alive and the standards are sustained is to remain privately held. In our case, we believe that having 100 percent ownership makes it even that much easier." Gary Erickson and Clif Bar's success provide an inspiring example. Yet, the many dilemmas of providing the six-billion-member human family with adequate nutrition still raise many questions and fierce debates. Wal-Mart says it wants to democratize organics and sell them cheaper. Consumer and organic groups say Wal-Mart will lower standards and end up outsourcing from China (*New York Times*, May 16, 2006) while *Business Week* covered "The Organic Myth" (Oct. 16, 2006).

Genetically modified organisms, or GMOs, have been heralded as the answer to poverty and malnutrition by industrial food companies. Most U.S. soybeans are now genetically modified, and Monsanto produces most of U.S. production of maize and soybeans. The United States produces 60 percent of the world's total of genetically-modified crops, while Argentina, Australia, Canada, and Brazil make up the rest. The European Union has long banned genetically-modified foods, due to consumer rejection of them. In 2006, the WTO ruled that the E.U. bans on genetically-modified foods were illegal, at the urging of the United States, as the Bush administration succumbed to lobbying by Monsanto, Aventis, DuPont, Dow Chemical, the National Corn Growers Association, and the American Enterprise Institute. Consumer and environmental opponents are still hopeful that the WTO rule changes will not change the widespread consumer rejection of genetically-modified foods in favor of continued high demand for more local organic foods (InterPress Service, Feb. 6, 2006). Fears over the health and environmental risks associated with these genetically modified foods have escalated the public debate.

Large-scale food manufacturers—GMO, organic, or otherwise—have not only impacted small-scale providers in the United States but have deeply affected farmers in developing nations such as Thailand where agriculture remains the primary source of livelihood. Nicola Bullard codirects Focus on The Global South, a nonprofit group based in Bangkok, that supports local economies and rural development. Nicola explains, "The important thing to know about Thailand is that more than 60 percent of the people are engaged in agricultural production. In the past twenty or twenty-five years there has been a real transformation in the agricultural sector. It's become very oriented towards exporting food, rather than growing food for local production. What this has meant is that agribusiness, and large-scale agriculture production has become the norm. Now the local farmers are really suffering the consequences of this, in many respects because they are losing a lot of autonomy. They're being pushed into producing food for exports. They have very little control over the market. They have no say over the price of what they grow. Obviously, if the price of rice, or the price of soybeans, or the price of onions fall, then the farmer still has the same cost for putting in the fertilizers, the labor they put in, the seeds, and so on and so forth. Many of them are much more in debt now than they were even twenty years ago. In the 1960s, 20 pecent of rural households had debt and now 80 percent of rural households have debt. So the farmers have really tried to organize themselves, to find solutions to this problem, to find ways to feed their families, to get good prices for their food, to resist the imposition of certain kinds of seeds, certain kinds of pesticides and fertilizers, and so on.

Nicola Bullard
Focus on the Global South

"In Thailand you really have fantastic movements of farmers organized in alternative production systems, creating local markets, re-capturing a lot of the old technologies, the diversity in the seed production, in the kinds of rice that they grow. They're really trying to carve out—not a direct challenge to the agribusiness—but they're trying to find a way not only to survive but to prosper."

In India, much of Asia, and Latin America, the story is similar. Indian activist and scientist Vandana Shiva (chapter 3) has also helped inspire farmers and consumers to action and is shaping public policy and corporate procedure in the process. Vandana explains why she founded Navdanya, based near Delhi, "The vicious cycle of industrialized, globalized agriculture is based on three very, very big myths. First that industrial agriculture with chemicals and genetically engineered seed produces more food. It does not. I spent fifteen years of my scientific life working on ecosystems and farming systems that are biodiverse, organic, ecological and they have tens to hundreds of times greater productivity—without using any external resources. The second big myth is that globalized agriculture is based on competition. No, it's based on five giant agri-business corporations controlling input, distribution, and the consumer price. The third very big myth is it's based on self-generating surpluses. Even the United States is not a food surplus country. It is importing more food than it's exporting. What the system has done is created what we call a great food swap. Everyone is exporting and everyone is importing. Only the trading agribusinesses are making a buck in exporting and importing. Their profits shoot up, but meantime the small farmer disappears. Consumer health is destroyed with one third of America suffering from the obesity epidemic.

"I think every one of us has an obligation to be a co-producer of a sustainable healthy, just, peaceful economy of food. The small farmer still accounts for the largest number of producers. But every one of us, even those who don't live on the land are co-producers because the moment we make a choice about what food we eat we are choosing the production system we support. In the very act of eating we are shaping the economy. In the very act of eating we decide whether the small farmer will survive or agribusiness will have more profits."

In the United States, Canada, Europe, and other mature industrial societies, there are millions of citizens and food consumers now supporting small farmers' movements with their purchases and political activism. Some lobby and picket the World Bank, the IMF, and the WTO, while others look for Fair Trade labels, such as those promoted by Forestrade, Inc. Thomas Fricke, cofounder, explains, "We're really creating a new model, which is, that doesn't fit into the neat boxes such as a broker, or a trader, or importer. We are really catalysts for what I call multi-stakeholder partnerships—farmers, entrepreneurs, co-operatives, warehouses, and a number of companies in both the United States and Europe that are our partners, that are helping to really get the story out, that are really helping to build the market." Anthony Rodale agrees, "The more that people get into growing food organically to meet the incredible demand prices will start to come down. How do we get healthier, fresher foods into the urban areas, into the rural areas where people's diets aren't always the best? It is by farmer's markets, by getting your local supermarket to have local foods, and also the direct connection with the farmer to the local consumer. There are abundant, diverse ways where we can show these examples of success around the country and internationally."

> We're really creating a new model, that doesn't fit into the neat boxes such as a broker, trader, or importer. We are catalysts for multistakeholder partnerships—farmers, entrepreneurs, co-operatives, warehouses, and companies in both the United States and Europe.

Fast food companies are in the crosshairs of consumers for serving salty, fatty meals, and contributing to the epidemic of obesity and diabetes. One fast food company, Panera Bread Company is thriving by serving fresh whole grain sandwiches, pizzas, and salads while paying employees well, offering health insurance, 401Ks, and discounted company stock. Ranked thirty-seventh in *Business Week*'s Hot Growth list of small companies, Panera earned $81.1 million profits in 2005 and its stock quintupled to $73 a share (*Business Week*, Apr. 17, 2006). Food and agriculture issues are now at the top of agendas at the Group of 8 Summit meetings and at the WTO. The Group of 20 spearheaded by Brazil, India, and China now demands equity and fairness for food producers in all developing countries (chapter 6).

Today, industrially processed food—whether canned, dried, frozen, or preserved—is losing favor among health conscious, upscale consumers. Even cooking food reduces its nutritional value somewhat. So many trendy restaurants now feature salad bars and raw natural organically grown foods that must be meticulously grown and cleaned. Many foods, including milk and fruit juices still need pasteurizing. Consumers are also looking for produce and foods uncontaminated with pesticides and toxics, like mercury, which concentrates in the food chain.

Britain's National Consumer Council offers a Health-Responsibility Index, which rates supermarkets to protect children from junk foods and growing childhood obesity. In 2006, for the first time in history U.S. children are sicker than the generation before them. Obesity affects nearly one-fifth, triple the prevalence in 1980; autism has increased one thousandfold in a generation; asthma is up 75 percent; life-threatening food allergies increased nearly sixfold; and nutrient deficiencies not seen for decades are now prevalent again, according to a report by Judy Converse, MPH, RD, LD, a licensed dietitian specializing in medical nutritional therapies for children. She and other health professionals note that increasingly aggressive vaccination—children are advised to get fifty-four vaccine doses (most containing mercury) by age twelve—may be causing health problems (www.redflagsdaily.com). Many parents in the United States have succeeded in ridding schools of junk foods and sugar-filled soft drink vending machines since evidence that they may contribute to attention deficit disorder and behavioral problems. The growth of Ritalin

Farmers Market

use is highly controversial in Britain, where prescriptions for children soared from four hundred thousand in 2000 to over seven hundred thousand in 2002, according to *The Economist* (Dec. 4, 2004), whose editors asked whether bad behavior might not be caused by bored and badly nourished children.

Organically grown produce cannot keep up with consumer demand, growing at 20 percent per year. The United States still has to import organics. The irony is that just as people know their health benefits, they must battle to get clear labels on other foods to know whether they contain GMOs. European consumers, unlike Americans, have so far been vigilant about GMO labeling. Laws there prevent sale of GMO-containing foods and many GMO crops produced in the United States are barred from export into Europe. A January 2005 report on Monsanto by Innovest Strategic Value Advisors pointed to the potential financial and reputational risks posed by its GM strategy. Monsanto paid $1.5 million in fines after the U.S. Securities and Exchange Commission (SEC) found the company had bribed an Indonesian environmental official over GM crops (www.socialinvest.org). Now, large GMO producers are also buying up seed companies—a cause for alarm since biodiversity is a key to sustainability. In today's globalized economy, all these changes in our food supply affect people in perverse and very different ways. Global fish catches are deteriorating and many large fish species face extinction. The growth of salmon farming worldwide is at the expense of other wild species. The patenting of native plants has provoked many national movements to protect them. In the United States, Europe, and Japan, contract agriculture is growing. Consumers form groups and contract with local farmers to grow customized organic crops. Luckily, healthy alternatives are now widely available in health food stores, supermarkets, restaurants, and your local farmers market. And don't forget to ask for them in conventional supermarkets as well. My friends Anna and Frances Lappe (author of *Diet for a Small Planet*, the 1970s classic) offers Ten Actions for Just Food and a Just World, from *Hope's Edge* (2002):

1. Enjoy food fresh from the farm. Buy directly from family farmers, look for family farm products, and encourage your local grocery stores and restaurants to do the same. To find local foods near you, visit www.localharvest.org and www.sustainabletable.org.

2. Vote your values with your dollar (and fork!). All of our consumption, savings, and charity choices make a huge impact. Find out where your bank, university, or pension invests and talk with them about choices that promote the health of workers and the planet. Learn more at www.socialinvest.org and get inspired by successful campaigns at www.ran.org.

3. Eat a sustainable and whole-foods diet. Support farmers raising produce and animals sustainably and in the process eschew the factory farming that contributes to air and water pollution, as well as global warming. Learn more about organic foods at www.organic consumers.org. Find meat raised sustainably at www.eatwellguide.org.

4. Support fair trade products and worker rights. Fair trade ensures farmers get a fair price. We can now buy fair trade coffee, tea, fruit, and more and bring fair trade into our local cafes and restaurants, hospitals, and schools. Find out more at www.transfairusa.org and get involved at www.globalexchange.org and www.tradematters.org.

5. Transform the buying power of your community. We are all part of institutions—churches, hospitals, workplaces, schools, city councils—that we can encourage to make purchases based on shared values. For instance, to find out more about bringing fresh, local, and organic foods into your school or other institution, visit www.foodsecurity.org.

6. Create "brand-free" zones. Advertisers spend billions every year to tell us what to eat, wear, and believe in—ads that bombard us in the classroom, doctor's office, even public bathrooms. We can create "brand-free" zones in our kitchens, schools, medicine cabinets, and more. Visit www.commercialfree.org and be inspired at www.adbusters.org.

7. Get a diverse media diet. Although six corporations control most of the major media, we can tap a vast, independent network for diverse information. See www.indymedia.org, www.gnn.tv, and www.frepress.net to get involved in bringing media democracy to life.

8. Get involved with the issues that matter to you. We can make our voices heard by joining advocacy groups, writing our elected officials, and getting involved with groups in our communities. Learn more about the issues that you care about, and find out how people are organizing to make a difference. To learn more about food, farming, and trade policy, visit www.foodfirst.org, www.publiccitizen.org, www.marketradefair.com and www.iatp.org.

9. Host a teach-in, study group, or gathering. See www.moveon.org for creative ideas about gatherings, events and local organizing around causes that matter to you and visit www.eatgrub.org for ideas about creating community-building dinner parties.

10. Vote! Clear and simple. Join www.indyvoter.org or other groups getting out the vote and building democracy—locally and nationally.

And, don't forget www.EthicalMarkets.com!

ROUNDTABLE
WALKING THE TALK

Simran Sethi and George Siemon

Simran hosted George Siemon, CEO of Organic Valley of LaFarge, Wisconsin, and Hewson Baltzell, our stakeholder analyst and co-founder of Innovest Strategic Value Advisors, to examine North America's largest organic farming co-operative. The company is the only organic brand solely owned and operated by organic farmers.

George Siemon: A co-op is much like any corporation except that it's built on serving its owners rather than serving them through building valuation.

Hewson Baltzell: George, how does the actual governance structure work in the co-op?

George: Well, of course, we're like any business. We have a corporate board consisting of all farmers. Besides that, our co-op has milk and eggs and juice and a wide variety of products and foods. Each of those groups has their own governance board. They all sit down and talk about all the issues that affect them whether it be quality or organic standards or supply management or their contracts, their agreements, and their pay price. So it really gives them a place to really engage in the business on a monthly basis. We send out very good minutes that describe the processes because we believe the educated farmer is the one that best will make it to the future."

Hewson:: You have something called the We The Farmers Pledge, so what exactly is that?

George: We have different standards that we have come up with. We have a pledge: a membership agreement and we have an affidavit process. We make sure farmers understand they're signing up for more than just a marketplace—they're joining something. We make them invest in our co-op as well and that really is a sign of commitment.

Simran Sethi: George, what do you do to ensure that the farmers are adhering to the organic standards?

George: The most important thing, I think, is just the peer pressure they get from their own farmers. So by having the regional reps in every region there's a lot of contact. If there are issues, we've got a person nearby that can check into it. Of course, rumors are a great integrity check!

Hewson: You also have some programs that help stabilize income. In the market, the price of milk moves up and down and that can be destabilizing for the family farm. Can you talk about that?

George: Sure, one of our biggest goals when we set out was to try to help farmers to a sustainable living and a stable income so they can plan their life. This is one of our greatest accomplishments and part of the whole organic dairy industry rewards. Farmers can plan on what they're going to get paid this year for their business purposes. Sure, there might be a little variation here or there but not anything noticeable.

Simran: And your regional distribution system allows people to purchase their goods locally, right?

George: The organic family products that they're purchasing are actually created by farmers that are in the region. People want to reward farmers for doing something right. Farmers are eager to be part of that and really want local production. So, if we're in the New England area, our milk is labeled New England Pasteurized, so farmers can take pride in having their milk at the local store. We have them go to demos or speak at trade shows to get them out there and put on their local face. This enables us to work with all the bigger chains and all the regulations and also to address the needs that people have in their region making that connection between the consumers and the farmers.

Simran: What is your relationship with chains? Would Organic Valley products be available in a place like Wal-Mart, which is getting more organic products?

George: The natural food trade is our natural home, but we certainly sell to the bigger chains as well.

Hewson: What about actual certification? I know there are debates about how rigorous organic is. Does that affect your business and do you have an opinion about that regulation?

George: Certification is a very valid system. We're now part of the USDA, which has given us some merit and recognition as a mature industry along with the headaches of working with government.

Hewson: What about the premium that customers have to pay. Let's say, a gallon of milk that's organic versus the commercial brand? How do you justify that to consumers?

George: Well, there's three issues: Organic feed's are quite expensive and in quite short supply, so there are additional costs. Two, we're trying to overcome what might be called a bankrupt food system of farmers going out of business every day. So we're trying to get those people up to a sustainable living. And three, we're still inefficient although we're quite proud of what we've achieved. Organic milk is 3 percent of the milk market now. That's a huge accomplishment compared to where we were ten years ago. We hope that our higher margins might decrease over time.

TWELVE

Health and Wellness

MOST ASPIRING DOCTORS ARE FAMILIAR WITH THE EDICT from the Greek physician and Western father of medicine Hippocrates to "first, do no harm." Judging by the state of medical care in the United States today, this simple requirement is not an easy one. Conventional health care in the United States has reached a crisis point—with widespread dissatisfaction among patients, doctors, nurses, hospitals, and over forty-six million people with no health insurance. "America's health system is a monster" declared *The Economist*—in its special report, "America's Healthcare Crisis" (Jan. 28, 2006), which pointed out that the U.S. health care system is by far the world's most expensive, at 16 percent of U.S. GDP—almost twice as much as the average in other industrial countries with no better outcomes. Many U.S. reformers, including Hillary and Bill Clinton during the Clinton administration, have been up against powerful lobbies like the American Hospital Association, the American Medical Association, big pharmaceutical companies, insurance industry, and manufacturers of high-tech medical equipment. The shame of the system is its unfairness, evident in those forty-six million ranks of the uninsured, a number that continues to creep up. *The New England Journal of Medicine*'s study in 2004, "Class: The Ignored Determinant of the Nation's Health," \ the truth: the poor die younger and are less healthy for obvious reasons, including substandard housing in dangerous neighborhoods delineated by race and class. Such social determinants of health start early in life, reported Dr. Nancy Krieger of Harvard's School of Public Health, and her brother James Krieger MD of Public Health Seattle.

The World Health Organization firmly acknowledges the link between socio-economic status and health—a relationship toward which the United States has turned a blind eye. As most other developed democracies switched to single-payer national insurance, the United States clung to its private-payer, employer-based mish-mash of providers and insurers, and its enormous bureaucratic, advertising, and marketing overhead while committing to costly state and federal programs under Medicare and Medicaid. Only the Veterans Administration, ranked as the best medical care in the United States, is run as a single-payer system (*Business Week*, July 17, 2006). As the debate about the U.S. medical system reaches a fever pitch, corporations are calling for change, dropping health coverage, and demanding government relief due to the staggering costs of employee health insurance. For example, General Motors' health bill for its employees and retirees adds $1,525 to the cost of every vehicle it produces in the United States.

As the debate about the U.S. medical system reaches a fever pitch, corporations are calling for change, dropping health coverage, and demanding government relief due to the staggering costs of employee health insurance.

The health experts at The Calvert-Henderson Quality of Life Indicators, which are regularly updated at www.calvert-henderson.com have been reporting on the wasteful, inequitable U.S. health system in their Health Indicator. This report notes that rising health care costs seem to be accepted by those who can afford them, because they assume that better health results from more expensive care.

In reality, Americans don't enjoy any better health than their counterparts in other industrialized nations—also reported in Dr. Barbara Starfield's studies at the Johns Hopkins School of Medicine in Baltimore, Maryland. Among thirteen industrialized nations—including Japan, Sweden, France, and Canada—the United States was ranked twelfth, based on the measurement of sixteen health indicators such as life expectancy, low-birth-weight averages, and infant mortality. *The Human Development Report* and its Human Development Index (HDI) finds similar results (chapter 1). Irrational elements in the U.S. system include the excessive attention on wealthy well people, cosmetic surgery, and the growth of "boutique" practices where doctors accept large annual retainers from well-off patients for home visits and lavish care, according to the *New York Times*, October 30, 2005. Many patients are overmedicated with expensive drugs of dubious value for mood enhancement, as described in *Trouble in Prozac Nation* (*Fortune*, Nov. 28, 2005). *The Baby Business*, by Debra L. Spar (Harvard Business School Press, Cambridge, Mass., 2006) describes the astounding growth of the fertility industry, now at $3 billion annually. Furthermore, Dr. David Eddy, pioneer of evidence-based medicine finds that only 20 to 25 percent of all treatments have been proven effective (*Business Week*, May 29, 2006).

Meanwhile, costs climb at over twice the rate of inflation, co-pays and deductibles skyrocket, and more employers' benefit packages shrink or disappear. Scared employees have been buying medical discount cards that don't deliver on their promises, according to *Business Week*, December 26, 2005. A radical approach, mandatory purchase of health insurance by all individuals and firms, was instituted in Massachusetts by Governor Mitt Romney in 2006. Libertarians objected, but many other states may follow suit. Yet, such mandatory payments to an already bloated, wasteful system may just increase overall costs without reform.

As the system threatens to implode of its own weight, some powerful antidotes have emerged. Simple remedies, such as precautions by hospitals that could save one hundred thousand needless deaths a year include better cleanliness and fewer errors, would make a significant difference according to *Newsweek*, December 12, 2005.

Dr. Steven Lawless has found a way to bridge our existing medical system to the future, taking the best of the old system and using computer technology to ensure safer treatments and medications and fewer hospital errors. Estimates find that as much as $80 billion could be saved annually by applying computer record keeping used in most other sectors of the U.S. economy. Dr. Lawless, who also holds an MBA, is chief knowledge officer at the Nemours Foundation and is pioneering such computer applications to drug prescriptions and hospital, doctor, and patient records. He states, "We can't have 98,000 people dying a year from what we do! We should have *nobody*. Part of my background is that I'm a pediatric intensivist so my primary clinical role is in the intensive care unit. So I have a passion for taking kids who are critically ill and making them better. Anywhere between 20 to 40 percent of prescriptions have errors in them, which had to be corrected by a pharmacist or a call back to the doctor—mostly due to doctors' illegible handwriting, but the content of that prescription also may not be accurate. The doctor may not know other medicines the patient's taking or may not know what the right dose should be. Is the patient taking some herbals or whatever? A lot of this information may be unknown. How many of those prescriptions actually have led to a patient problem? On top of that, a lot of patients take the drugs and then another doctor or another system sees them. So, that lack of linkage, of tallying any adverse drug reactions may never actually get back to the doctor—or into the patient's record."

Doctors and hospitals have notably lagged behind most other sectors of the U.S. economy in using computerized information, according to a report in *The Economist*, April 30, 2005. Steven Lawless, of Nemours Children's Clinics, explains, "We are one of the nation's largest pediatric sub-specialty healthcare clinics in the country. We have electronic systems. We have a lot more integrated processes where business intersects with operations that intersect with quality and the clinicians. And one of my jobs with that is to bring it all together, to get them all to communicate together. A normal hospital gives about a million drug dosages a year. If you rack up the numbers, you're talking about maybe a hundreds deaths in a hospital a year. Some of these hospitals deaths could be resulting from medication errors. That could be solved if you have electronic systems that help prescribe, check, make sure everything's adequately done. They can assure that the right patient is getting the right dose so that harm to patients can be reduced. What we're trying to do is link up our systems so we're actually partnering with the pharmacist. They tell us what's going on. Everything's electronically recorded. They tell us if there's an error. We record it back to them. We're feeding all this back to our doctors now with a continual loop to let them know what's going on. There's a new social responsibility here. And that new social responsibility is if everything is getting linked, I know my record here is going to last for a long time! This information about doctors' records and competence can go filtering all over the place." Civic groups in many states maintain similar records on physicians, their competence, and the prevalence of malpractice suits against them. A good source of information is the Washington-based Center for Science in the Public Interest, publishers of *Nutrition Action*.

Increasing numbers of Americans are seeking out alternative providers and remedies rather than visiting conventional medical doctors.

A classic case of bureaucratic inefficiency is the 2004 Drug Benefit law, which began in 2006. Senior citizens were caught between dozens of insurance company plans, and state and federal bureaucracies—with few complete records of their medications. This boondoggle law, which funnels additional millions to pharmaceutical companies—big campaign contributors to both political parties—is one example of how costs spiral out of control.

Unsurprisingly, people are turning toward less costly healthcare options for help. Americans now average $40 billion annually on alternative health care and on vitamins and supplements. The burgeoning new sectors of our economy based on preventive, natural, and systemic approaches to wellness are redefining health for all. Increasing numbers of Americans are seeking out alternative providers and remedies rather than visiting conventional medical doctors and their facilities. Books by alternative MDs, including Dr. Deepak Chopra's *Ageless Body Timeless Mind* (1993) and Dr. Andrew Weil's *Spontaneous Healing* (1995) are perennial best sellers.

The New York Times reported on February 6, 2006, about the ways in which patient's have lost trust in what many call the medical-industrial complex, and why so many prefer alternative, complementary providers, many of which are based in healing traditions that have been in existence for centuries such as homeopathy, Ayurveda, and traditional Chinese medicine. The almost excessive specialization that afflicts our current medical approach is part of a three-hundred-year academic tradition of reductionism: the idea that wholes can be understood by dissecting them and studying their parts. This so-called Cartesian approach—named after the French mathematician Rene Descartes—led to an explosion of human scientific knowledge. Now, we are beginning to put these pieces back together in a larger worldview and more systemic approaches, which I have termed the Post-Cartesian Scientific Worldview (Henderson, 1981). We are slowly remembering that our bodies, minds, attitudes, spiritual lives, social circumstances, and physical environments interact to keep us well, or make us sick. Many Americans visit complementary healthcare providers rather than traditional mainstream doctors because they feel that conventional medicine doesn't adequately address their needs. Others focus on exercise and account for the burgeoning fitness industry. One pragmatic movement is that of office workers who skip elevators and walk up the stairs. In the recent Empire State Building Run-up, an annual race, contestants dashed up eighty-six floors in between ten and twenty minutes. Such races are now held in office buildings in many countries (www.towerrunning.com).

Former Rhode Island Congresswoman Claudine Schneider sponsored the "Global Warming Prevention Act of 1988," the first in Congress, and led environmental legislation in the 1990s. She is now a senior consultant with the renewable energy firm, Econergy International, in Colorado. Claudine recounts her personal story. "There are many things that doctors don't tell you. First of all, they don't emphasize how important it is to look at your own emotional state of mind, to listen to your body, to pay attention to it. They also do not emphasize the importance of getting several opinions. And few doctors realize there are different alternative approaches. When I had cancer, a second time, about eight years ago, it was a very low point in my life. The doctors said, *We're sorry, there's nothing we can do.* Fortunately, I never was really too inclined to take a doctor's opinion, unless I had gotten several other opinions. What was most important is I made an emotional turnaround and decided I was not ready to go yet. So, I spent a whole year researching what I should do even though this did look like the end. I made a lot of phone calls, I read a lot of books, and I found my way to a clinic in the Bahamas that stimulates your immune system so it can fight the cancer cells. I also went to Dr. Nick Gonzales in New York City, one of the first doctors to receive a major grant from the National Cancer Institute for using alternative approaches to dealing with cancer. His specialty is pancreatic cancer, which usually entails a three-month life expectancy. His patients have lived three to eleven years. When you're looking at alternative approaches you make sure they are science-based. My doctor in New York provides vitamins and enzymes and a certain diet to follow. When I first had cancer, I became

Claudine Schneider
Former Congresswoman

a vegetarian and stayed a vegetarian for twenty-five years. But he said, *Oh no, you have to be eating red meat!* I thought, oh my God, *I don't know if I can do this!* But I did, I mean if that's what it means to stay alive, life is too great and it is an incredible gift."

Claudine, again an active, healthy woman, worked through Congress to persuade the National Institutes of Health to begin researching alternative therapies. She was effective in changing the direction of other agencies as well. "One of the things that Senator Tom Harkin and Congressman Berkley Bedell and I had all fought for was to make sure that we had an Office of Alternative Medicine at the National Cancer Institute. That office is there today and does the research on some of the treatments that are perceived as being bizarre and unusual. Some of them work. Some of them absolutely do not. What's interesting is that some work for some people, and others do not—the same way that chemotherapy works for some people and for other folks it does not. So there is no silver bullet. We have to really pay attention to what we think is working."

Dr. Jim Gordon founded the Center for Mind-Body Medicine in Washington, D.C. "The Center for Mind-Body Medicine is a non-profit educational organization here in Washington, D.C. What that means practically is putting a major emphasis on self-awareness and self-care. We look at every aspect of healthcare: from delivering services to people with serious illness to medical education; the education of other health professionals to using self-awareness and self-care; and mutual help as a basis for working with war-traumatized populations. I was asked by President Clinton to chair the White House Commission on Complementary and Alternative Medicine Policy. Initially, medicine is very conservative and all new ideas are challenged, with considerable energy. That's quite appropriate, because medicine is really working with issues that are so fundamental to all of our lives. It's also appropriate that all new theories and all new practices in medicine be subjected to close examination.

"What we call alternative medicine is really just everything that doctors over the age of thirty did not learn in medical school. Many of these therapies were also very strange to me, in the beginning. Acupuncture, meditation, herbal therapies, and dietary therapies I'd never heard of and they never had been mentioned when I was at Harvard Medical School. I'd never seen them referenced in any of the medical journals. I got interested because I was just curious about these other traditions. I thought that if people have been using these approaches in India or China for thousands of years, that there might be something to them. So I just began to look in a kind of open-minded way. I think that some in the medical establishment shared that interest in exploring. Yet, many more were just kind of put off and thought, *Well, if I didn't learn it in medical school and residency training, and it's not in the journals I read, then it's probably not important,* which, unfortunately, is less scientific and more arrogant than we should be.

Dr. Jim Gordon
Center for Mind-Body Medicine

"The use of one very simple therapy—massage for premature infants—has been shown by a wonderful researcher, Tiffany Field, as able not only to help premature infants grow and develop faster, but it also helps to ensure that they'll have fewer complications. These babies will be able to leave the hospital on average of six days earlier than infants who weren't massaged. This was a very careful, randomized, controlled study and showed that we would save up to $10,000 for each of these infants. So the question arises, if this is not only good for their health, if it's simple to bring into a hospital, why isn't every hospital in this country offering massage to its premature infants and perhaps to its full-term infants as well?"

In Ithaca, New York, Paul Glover, inventor of the famed local currency, Ithaca Hours (chapter 3), also spearheaded Ithaca's community health co-operative, the Ithaca Health Alliance, which provides basic health care for its members for $100 per year. The program has grown over the past ten years and now has over six hundred members and some 135 licensed providers (www.ithacahealth.org). Paul is now in Philadelphia organizing a similar group, PhilaHealthia, all described in his forthcoming book, *Health Democracy* (www.health democracy.org).

Health, as we know is related to diet and lifestyle habits. But it's also tied to environmental conditions such as the quality of the air we breathe, the water we drink, and the products we consume. For example, the wide use of asbestos has led to illnesses and deaths along with the bankruptcy of many companies, with a projected $140 billion fund for these victims still contested between states (*Business Week*, Mar. 6, 2006). Jeffrey Hollender, founder and CEO of Seventh Generation Company, which makes nontoxic home products from toilet tissue to cleaning and laundry items, was determined to remove household hazards. "We think about the food we eat. We think about the products that we use on our skin. The air we breathe on average in our home is two to five times more polluted than the air outside of our homes, according to the EPA. Indoor air quality is a critical issue. We can look at a specific situation: when you are using a household-cleaning product, and you can smell it. Normally what you smell, whether it's a scent or a chemical odor, is what is called the VOC, a volatile organic compound. Those VOCs come out of cleaning products into our air. We know without question that those VOCs are asthma triggers for children. They also

Jeffrey Hollender
Seventh Generation Company

cause allergies and various types of irritation to our lungs. We have to take responsibility not just for the food and the products we put on our skin, but very importantly, for the air we breathe. The most effective way we can do that is by making sure that the products you bring in your home don't emit VOCs that pollute your air." Jeff's products are no longer just in health food stores—you can find them on the shelves of many mainstream retailers.

Understanding we exist within a larger system, we know that the choices we make do impact our health, our environment, and contribute to our healing processes. The Seva Foundation founded by physician and high-tech entrepreneur Larry Brilliant, MD, MPH, has taken a global approach to healing, working to support communities by treating individuals. Larry recalls why he founded the Seva Foundation, based in California, but working worldwide. "Seva was started by my wife and I and a group of our friends, many of whom had worked in the smallpox program. We'd been health diplomats working for the World Health Organization or university professors. After we eradicated smallpox, we wanted to do something else like that again. To see a disease eradicated—and smallpox was the first and so far only disease in history to be eradicated—it's like a mountain climber climbing a mountain. It's such a preposterous achievement that we wanted to do it again! So we looked around for other diseases, conditions that were amenable, like smallpox was, to a worldwide campaign. Blindness can't be eradicated because every year there're new cases of blindness. Yet, blindness is something that seemed to be such a poignant affliction and we had also seen so many blind people blinded by smallpox, and that is a double affliction.

Larry Brilliant with Friends in Nepal
Seva Foundation

"I think it was natural for us to come together to try to form an organization that could do some of the things the World Health Organization or UNICEF can do, but as a private, not-for-profit, tax-exempt organization. We've been at it for twenty-five years. We're still working in all the same villages, all the same countries. The difference is instead of sending in American volunteers, Nepal, for example, is now exporting ophthalmologists who've been trained. That program is now economically and emotionally self-reliant, using that money to train other people in eye care. In India, Dr. Venkataswamy has created the Aravind Eye Hospital—beginning with fifteen beds in his home. It is now the largest eye hospital in the world! He has seven hospitals. Last year, they did almost three hundred thousand sight-restoring operations. In order to make it possible to give back sight to so many people, they now manufacture their own inter-ocular eye lenses, the implants you put in people's eyes after you take out their cataract lens. Last year they made almost a million of those."

This inspiring work is a good example of C. K. Pralahad's Bottom of the Pyramid approach, mentioned earlier in this book, to serving the world's two billion people who subsist on less than $2 a day. The R&D approach is all about effecting huge cost-reductions, for example, learning how to mass-produce hugely expensive crafted surgical instruments. Larry Brilliant now heads Google's foundation.

Another novel approach to human health involves bio-electrical fields. Dr. Beverly Rubik, president of the Institute for Frontier Science in California, explains. "There's more acceptance and a little more funding year by year for studying complementary and alternative medicine. We study the effects of extremely low-level electromagnetic fields for stimulating bone healing and many other uses. Innovation in this field is a very slow process. On the one hand because the funding is miniscule and on the other hand, there is a long process in getting these therapies that involve medical devices through the U.S. Food and Drug Administration (FDA) process. The FDA is largely set up for drugs, three levels of clinical trials and a lot of safety issues. On the other hand, these devices operate usually with extremely low-level energies which conventional science says shouldn't even have a biological effect. So they're hardly harmful! Yet, we have to go through the same laborious and very expensive clinical trials of three phases to prove our point and get FDA acceptance. So there's a long road between doing the research and getting FDA clearance. Even then it takes more time to become accepted in the mainstream medicine. For example, the electromagnetic therapy for bones, stimulating fractures that have not healed well, has been accepted by the FDA for over twenty years, but in fact it's still not widely used today in medicine." Much of this work is based on the pioneering research of Dr. Robert O. Becker in *Cross Currents* (1990) and other books, which led to the use of microcurrent stimulation for pain relief.

So, as we have shown, a lot needs to change in our U.S. healthcare system—and much is changing. The time seems ripe for a wholesale revamping of our current approaches. For example, the cover of *U.S. News and World Report* (Jan. 31–Feb. 7, 2005) screamed, "Who needs doctors?" Their answer was: Your future physician might not be an MD, and you may be better off. A 2005 Harvard study found that medical bills are a leading cause of half of all personal bankruptcies in the United States. Many of the new providers featured in this chapter are becoming mainstream. In addition:

- Births assisted by midwives and private birth centers are thriving from coast to coast.
- Hospices, caring for those who wish to die in peace are now an accepted alternative.
- Nurses, woefully underpaid shoulder greater burdens and bigger roles in hospitals. *Newsweek* reported on the national shortage in their December 12, 2005 issue.
- Nurses provide a wide array of medical procedures formerly performed by MDs and half a million more are needed.

The 2005 debate over Social Security has obscured a real budget crisis: the Medicare Trust Fund is projected to be empty by 2019—partly due to the huge cost of drug benefits. The 2004 law prevents lowering costs by using the government's bulk purchasing power or outsourcing from Canada. If these trends continue, U.S. companies may join those groups calling for single-payer universal health insurance! Can malpractice costs be lowered? Yes! By raising insurance rates for only those few doctors who incur the most lawsuits, and many hospitals are now offering free group coverage to recruit more doctors.

With U.S. health costs at 16 percent of GDP and still climbing, it's no wonder that less costly, less invasive, often less dangerous medical approaches are winning out. These young companies and alternative providers are stressing prevention, fitness, and healthy lifestyles and have burgeoned into a multibillion dollar sector of our economy. No wonder it now tops the political agenda of politicians across the spectrum—as well as in the priorities of U.S. citizens. Corporate America is getting activated. An eighteen-page advertising supplement in *Fortune*, October 16, 2006 promotes wellness programs for employees as cost-saving measures. But more radical approaches will be needed to address the systemic waste and inequities in U.S. health care.

ROUNDTABLE
WALKING THE TALK

Paul Freundlich, Simran Sethi, and Dr. Barbara Glickstein

Simran hosted Dr. Barbara Glickstein, director of Clinical Programming and Community Outreach at the Continuum Center for Health and Healing at Beth Israel Hospital in New York City, and our stakeholder analyst, Paul Freundlich, president of the Fair Trade Foundation to learn more about this largest and most comprehensive academic center for integrative medicine in the country and a leader in this field. Ethical Markets Research Advisory Board Member Rena Shulsky, founder of Green Seal labels for environmentally friendly products, also helped launch Beth Israel's Center and ensured that its facility incorporated non-toxic furnishings and building materials.

Barbara Glickstein: The integrative approach to medicine is an approach that takes the best of allopathic or conventional medicine. We look at the broad range of ancient healing traditions, whether it's Aruveydic, indigenous healing practices from the native community, mind-body medicine, East Asian medicine, and others. We look at the best of those approaches based on evidence that we have. We then look at each person and as they come for healthcare, we try—with them—to make the best medical decisions and healthcare decisions for them to maximize and optimize their health.

Paul Freundlich: This sounds like it was very adventurous for Beth Israel to launch your operation and to support it?

Barbara: Well, traditionally Beth Israel Medical Center has been a leader in many ways. We're an incredibly diverse hospital with people from all over the world coming to see us, and people from all over the world are working there. So, in fact, we have a population of people that represent these indigenous healing practices, this is their primary care medicine. We're all highly credentialed, we're all seasoned practitioners—people who have come to the Center are highly trained and highly experienced in their practice.

Simran Sethi: This feels like an approach where patients are just as invested in this sort of diagnosis and the treatment as the practitioners are?

Barbara: If you asked me what the biggest cultural shift about integrative medicine is—this shift is that it really does call upon each person who's entering the door to make decisions about their health. We try and, I think it's critical to meet patients where they are. In other words, you can't change your entire diet, start exercising, and move if your home is a place that's perhaps not as healthy as it could be. So we work with people where they are financially as well. Some of our services are not covered by insurance—something we'd love to see change—but if someone needs to work within their resources, that's part of their health.

Paul: I was wondering whether your staff finds there are times where you sit down and take a look at individual cases or the general case load and see what's developing. Out of this whole range of traditional and alternative methodologies, what is the best approach to this patient or illness?

Barbara: We have a Fellowship in Integrative Medicine. We have weekly case studies where all of the clinical staff comes together and a fellow would present a case and the best practice would be discussed. Do I wish we could do that with every patient, every day? That would be ideal and a world that you and I would love to live in! But, financially that's not a possibility. What the discussions do is allow people to hear each other. Whether it is our clinical psychologist in Mind-Body Practice and a very fine analyst who says, have you thought of this question? Or whether it was our East Asian Chinese Medicine practitioner saying, let me teach you how to read the tongue and do a pulse, the next time they come in, they see if this is going on, or refer to it. The dialogue is the richness that's going on in our practice. That is why these seasoned practitioners who don't need another place to practice will come forth to work with us and work together as a team.

Paul: Does that extend into the mainstream practice of Beth Israel itself?

Barbara: We have a monthly Grand Rounds to which the entire hospital community and the entire systemwide community is invited. We have departments that have become more aligned in making referrals more regularly. I must say that there are often cases of patients who have chronic issues. To the credit of those departments and providers, they know at this moment in time, this is all they can do. They admit that with the medicine they have available, they've done their best. Sometimes, even those leading forms of medicine they're practicing and this expertise doesn't address chronic issues.

Paul: Is the health care that you're practicing more efficient and have you been able to gather the evidence for that?

Barbara: I wish I could say that we have a model in this healthcare system and in integrative medicine that's currently looking at the financial efficacy of these practices. I know it's a very difficult question. What I can tell you is that the more people take responsibility the better the outcome—I don't mean that as a moral point or finger-pointing I do believe that we have a culture, in some ways, of dependency on the sickness and medical models. These approaches have been cocreated by the population of those who seek care and the population of those who have been delivering and developing policies around health care. I think we need to look at ways to ameliorate unhealthy conditions: the environment, stress, where people work, where people are given day-care support, where people are given work leave that is related to raising families and taking care of our elders. These approaches would begin to shape health care into a very different and more effective model.

THIRTEEN
The Future of Socially Responsible Investing

IN CHAPTERS 1, 2, AND 9, we introduced some of the early pioneers of U.S. socially responsible investing, or SRI. Three pillars of SRI are 1) social, environmental, and ethical auditing, or screening; 2) community investing; and 3) shareholder activism. The fourth pillar is socially responsible venture capital, which is vital in seeding all the new companies needed in the shift to global sustainability. Total U.S. investments in SRI represent some $2.3 trillion and the movement has caught on in Canada and Europe. SRI is now spreading to Australia, New Zealand, Japan, China, and Brazil. Companies chosen after careful social, environmental, and ethical auditing also outperformed the S&P 500, Brazil's BOVESPA, and many other conventional indices—demonstrating that you can do well by doing good. On April 27, 2006, representatives of more than twenty pension funds from sixteen countries managing over $2 trillion of assets made a historic announcement at the New York Stock Exchange: their launch of Principles for Responsible Investment. In a matter of weeks, new signatories swelled the total to over $5 trillion. Denise

Nappier, treasurer of the state of Connecticut, a longtime leader in SRI noted "We are proud to endorse the principles, which recognize that social and environmental issues can be material to the financial outlook of a company and therefore to the value of our shares in that company." (More at www.unpri.org and www.unepfi.org.)

In this book and its companion TV series, Ethical Markets, we covered the many companies and CEOs creating higher standards and benchmarks for good corporate citizenship in the twenty-first century. The future of this movement for more ethical markets globally promises to be bright. We launched in Brazil as "Mercado Etico TV." Key drivers of this evolution of capitalism are civic society watchdog groups, aware consumers, employees, and citizens. They are augmented by the growing financial power wielded by socially responsible investors. We learned how this new force in capital markets steers companies toward higher social, environmental, and ethical performance, and fosters a new kind of corporation committed to all its stakeholders, not only its stockholders. Among the many new indexes we mention is that covering Lifestyles of Health and Sustainability Companies, the LOHAS Index (www.Lohas.com).

Launch of Principles for Responsible Investment

In 2006, venture capital began to surge into clean technology and green start-up companies as never before. Nick Parker, founder and chairman of Cleantech Capital Group, based in Toronto and Brighton, Michigan, has led this venture field for decades. Cleantech's regular Venture Forum in San Francisco, March 2006, attracted over five hundred venture investors, including Silicon Valley leaders like Vinod Khosla, who made his billions in AOL, Amazon, Compaq, Sun Microsystems, and Google. Dr. Zhengrong Shi, CEO of Shanghai-based Suntech Power Holdings Company, which had the largest initial public offering (IPO) in 2005 of $5.5 billion, and is China's largest provider of solar energy, also attended the Cleantech

Venture Forum. Bill Joy, cofounder of Sun Microsystems is another convert, now a partner in Kleiner Perkins Cauldfield and Byer, a top venture firm now part of the copy-cat Greentech Innovation Network.

Nicholas Parker
Cleantech Capital Group

Vinod Khosla is now dedicated to green technology and making petroleum obsolete with his campaign with Hollywood producer Stephen Bing, Californians for Clean Energy. An engineer from India's Institute of Technology, Khosla favors cellulose-derived ethanol, which has helped Brazil to become self-sufficient in energy, rather than auto fuels made from food crops like corn. Many environmentalists, including Lester Brown, founder of Worldwatch Institute (on whose board I served from 1975 until 2002), agree and also want all cars to be like those manufactured in Brazil: flexible-fuel vehicles (FFVs), which can use any available fuel. This FFV switch costs about $100 per vehicle and can provide the swiftest transition to hydrogen—meshing with Toyota and Honda's hybrid technologies, which can also switch to hydrogen. Since transportation uses some 50 percent of U.S. oil imports, Khosla believes this sector is the place to start. Oil companies now threaten to kill biofuels by dropping the price of oil. They still receive oil depletion allowances of 27 percent, instead of taxes on the pollution and damage they cause. Even at $60 to $70 a barrel, oil in the United States is still cheaper—when corrected for inflation—than when OPEC quadrupled its price in 1973—and about half the world price paid in Europe and Japan. Californians for Clean Energy asks Californians to increase taxes on oil production by up to $380 billion annually anytime oil prices drop, so as to protect and augment investments in clean energy.

A keynote speaker, Mark Donohue, general partner of Expansion Capital Partners of San Francisco and New York, urges his colleagues to go beyond traditional venture capital goals of financial gain and the exit strategy of going public on Wall Street with IPOs. Mark cautioned that the health of the planet itself should be the goal and using triple bottom line accounting should be standard in all Cleantech investments, as it is in his firm. Often, when such highly ethical companies go public, they fall victim to Wall Street's growth-at-all-costs analysts, which led to the wave of CEOs cooking their books to please them. Mark serves on the Ethical Markets Research Advisory Board.

Many of the technologies of sustainable development have been waiting in the wings for many decades: solar, wind, ocean, and biomass energy, eco-efficiency, recycling, better storage methods, fuel cells, and hydrogen. But corporations of the fossil-fuel age, largely for reasons of self-interest, have not invested in developing technologies that would be highly disruptive and eventually succeed them. So venture and angel investors are needed—along with a repeal of government subsidies to coal, oil, gas, and nuclear energy providers. A design revolution is needed to revamp economic models, our sprawling infrastructure, oil-dependent transportation, agriculture, chemicals, construction, and our wasteful buildings. Naturally, such systemic change requires decades and has been happening unobserved by media. Venture capital has always played a unique role in the U.S. economy—spawning our rapid technological innovation during the industrial age. Venture capital fueled the dot-com companies of the 1990s—and investors lost trillions of dollars when this bubble burst in 2000. After languishing and licking their wounds, venture investors have now discovered the next big thing—technologies needed to create more sustainable societies.

Robert Shaw, president of Arête Corporation, a venture fund for renewable energy, has been a leader in identifying and supporting alternative energy options. "Arête is a venture capital fund manager. We make investments in a lot of different sustainable energy technologies, small-scale technologies that try to provide the electric energy that the world needs and to do it in an environmentally benign way, and locally. We like the *Small Is Beautiful* idea [referring to E. F. Schumacher's 1973 book]. If it's close to you and you know how to control it and run it, it feels better and it's less out of your control. That's what we're doing. We're making investments just like any other venture capitalist, but in this particular space. Venture capitalists take other people's money and invest it in companies in order to make money for those folks and for ourselves as venture managers as well, because we take a small piece of the game. That's our incentive for trying to do well. It's a very high-risk field because we're swinging for the fence every time and a lot of the companies we invest in don't work. But some of them do work very well, and that's where the return comes from. So it's an exciting business; it's not like putting money into U.S. Treasury bills or even into the stock market because a lot of the things you try are going to fail. The excitement comes on the things that work! We've built some beautiful, wonderful companies: Evergreen Solar Corporation, Distributive Energy Systems Corporation, to name a couple, American Superconductor, Ballard Power, which is in the fuel cell business; Cell Tech Power is another company that created a very novel solid oxide fuel cell system. So when you have a few successes you make money for your investors! And you enjoy what you do!

Robert Shaw
Arête Corporation

"There are companies all over the planet working on these technologies. The United States is not alone in this race for sustainability. Europe and Japan, and now China, are making many of the same kinds of systems. The Chinese market is probably the single most attractive market, because the Chinese are growing at an enormous rate. They don't have anywhere near the amount of electricity they need to serve that growing economy and they would like to do it in a way that's not going to pollute the planet. The Germans, the French, the British, the Japanese, and the Scandinavian companies are all working vigorously on these technologies. So, it's going to be a horse race. I don't think we can say that any particular nation is likely to win at this point, but I think it's important that everybody be in the game, because in the end, it's the planet that we're trying to save."

A few years ago, such sentiments were unheard of on Wall Street or dismissed as the wishful thinking of hippies or tree-huggers. Not any more. The herd behavior of investors has initiated a rush, which may be too much, according to Nick Parker, who doesn't want to see a new bubble occur in cleantech. Only recently have behavioral scientists forced economists to question their belief that investors are rational and that they weigh all available information in pricing stocks and all goods. Today, the exploding of this myth by behaviorists has undermined this central tenet of economic modeling—however fancy the math and computer programs, as described in my "Twenty-first Century Strategies for Sustainability" (*Foresight*, Cambridge, U.K. Feb. 2006, downloadable from www.hazelhenderson.com.).

Economists, Wall Street analysts, and financial media have discounted the trend toward sustainable economies and socially responsible investing for two decades. Now, a tipping point has been reached in the growth of SRI and its robust rates of return. In the 1980s, SRI assets totaled about $40 billion, mostly owned by mainstream religious groups and charitable foundations whose values always went beyond the bottom line. Today, the Social Investment Forum reports that values-based investors account for more than 11 percent of all investments under professional management.

The global SRI movement is based on rigorous research, screening companies for social, environmental, and ethical practices including employment policies, human rights, consumer protection, and environmental stewardship—just as we do for Ethical Markets TV series. Innovest Strategic Value Advisors, a leading firm in this new branch of the auditing profession, researches such broader corporate performance. They issue their corporate audits to alert their clients (mutual funds, banks, pension funds, endowments) of the potential environmental and reputational risks such bad corporate behavior poses to their shareholders in damaging brand identity and stock prices. Such ethical auditing firms—now a burgeoning industry—often release the same kind of "Buy, Sell, or Hold" advice as traditional security analysts. As mentioned, the world's newest superpower "public opinion" has shown how easily corporate reputations and their stock prices can suffer from short-sighted, antisocial decisions.

Wayne Silby, founder and chair of the Calvert Group, the largest U.S.-based SRI firm, describes how these values-based investments drive businesses toward greater corporate social responsibility. Wayne remembers the early days, "When we invest our money, it's like voting for the kind of world we want to create. It's expressing our values. Do we want a company that believes in diversity, in terms of the values in our society? Do we want companies that have no regard for how they do their ethical drug trials in developing countries? Where is that responsibility? When you, as an investor, have that ability to have a say, you have a responsibility to exercise that say. So, the movement is really about joining together to express our values, and make sure that money makes the world we want. That change involves values."

The Calvert Social Investment Fund was founded in 1982 with a few million dollars and has grown to $10 billion in a whole family of funds today. (I served on its Advisory Council from 1982 until 2005.) Wayne adds, "What I'm most proud of is the fact that I feel we've provided leadership at Calvert. That isn't about us, it's more about people working together and we have done a number of innovations. Of course, we were early, so we had the chance and we had the money and had the resources to do it. Do you have children? Do you have grandchildren? I know that future generations are going to be left a legacy with regard to what we do now. They need air that they can breathe and a planet that has a kind of a climate that's livable—like we found it. I think it's a pretty basic principle in most religions or most attitudes that at least, clean up your mess, and don't leave the world a worse place." Wayne's venture investing—beyond his leadership in Calvert mutual funds—has helped many young start-up companies in the growing sustainability sector. Wayne cofounded the Social Venture Network and is active in Investors Circle founded by Susan Davis of Capital Missions (chapter 7) and other social venture groups.

Robert Monks, author of *The New Global Investors* (1998) and formerly responsible for pension fund oversight in President Reagan's administration, is another pioneer. "In my view, SRI is investing in companies and changing them. A socially responsible investor is someone who says to a manager doing an act that it is harmful to society, It's my company. I want you to stop doing this. I want you to do it in a way that is congenial to society. That to me is socially responsible investing. We've experimented over the course of the last two hundred years with a whole bunch of ways of creating wealth, and it turns out that the combination of the ability for people to invest money with limited liability, the capacity to hire people with particular skills and management, then the ability of the owners at any given time to sell their interests so they can diversify to other things has created a framework by which wealth has been created to an extent never before experienced."

Robert A. G. Monks
Author, *The New Global Investors*

Bob, like Wayne Silby and Nick Parker, is a member of the Ethical Market's Research Advisory Board. Bob is also a lead investor in the British firm, Truecost, which calculates social and environmental costs of production externalized from corporate balance sheets and passed on to taxpayers, future generations, and the environment. The successful Dutch computer systems integrator founded by

Eckart Wintzen, BSO Origin, was the first company to acknowledge these external costs imposed on society in its 1990 Annual Report. In citing the costs of its water waste, CO_2 emissions, and fossil-fuels impact on climate change, Eckart called for the now popular tax-shifting from income and payrolls to pollution, waste, and resource-extraction: Value-Extracted Taxes (VET) instead of Europe's Value Added Taxes (VAT). Eckart, now a green venture capitalist is also an advisor to Ethical Markets.

Susan Davis (chapter 7), president of Capital Missions Company in Elkhorn, Wisconsin, has been a social investor all her life. Susan is also a missionary in promoting all kinds of initiatives to expand socially concerned capital markets. While a vice president of the Chicago-based Harris Trust, she organized the Committee of Two Hundred—bringing together the two hundred highest net-worth women business leaders and entrepreneurs. The Investor's Circle she founded, is now a group of several hundred venture capitalists dedicated to financing companies contributing to a more sustainable world. "Actually, the socially responsible industry is thriving and has been outperforming its competitors. One example is the Triple Bottom Line simulation on the Web site www.capitalmissions.com where some of the leading investors in the country invested $100 million portfolios 100 percent in social investments. Those investments have been outperforming their financial benchmarks. All this is available free. This information shows the choice of investment products in each asset class from equity, international equity, fixed income, cash, and alternative investments, including venture capital. You can actually track the ten-year performance of more than thirty-one investment products that are in this Triple Bottom Line Simulation. You can see the asset allocation strategy that the treasurers who created the simulation actually used for their own large portfolios. So, it's a model that anyone can use and it educates any financier about social investing in a fun way."

Susan Davis also finds more financing is available. "The long-term investors who are looking for the larger return long-term, are jumping into solar. In Europe, I know a number of new funds from wealthy families. It's the same thing here in the United States as well. The Silicon Valley investors, who are at the top of the heap there, made a huge amount of money in that Internet experience. Many of those people had a social conscience. They wanted to do something that had global impact. They see renewable energy as a chance to do that now. So when recently the head of a major renewable venture capital fund spoke to the Executive Club in San Francisco, they had standing room only, the largest turnout they'd ever had." Susan is still forging ahead with her Solar Energy Circle of investors and her Tipping Point Network of pioneers in sustainability and also advises Ethical Markets.

Socially responsible investments, as we learn, are about good risk management. Since consumer trust and the stock markets have been shaken by corporate scandals, investors have switched into SRI mutual funds seeking ethical standards as well as stable returns on their investments. The persistent corporate scandals will be with us until global standards and greater transparency can steer twenty-first century markets toward urgent human needs and the new goals of sustainable human development. In *The United Nations: Policy and Financing Alternatives* (1995, 1996), which I coedited with Inge Kaul, pioneer of the U.N.'s Human Development Index (HDI) and Harlan Cleveland, former U.S. ambassador to NATO, focused on the U.N.'s role as a global norm-setter. The United Nations has fostered most of the global treaties on human rights, workplace standards, environmental protection, and all the now globally ratified priorities of health, education, poverty reduction, protection of children and refugees—the framework of multilateral cooperation that protects all the world's peoples. In 2005, the United Nations Global Compact added anticorruption to its ten good corporate citizenship principles, described in chapter 2 and the Millennium Development Goals. Much U.N.-bashing in the United

Much U.N.-bashing in the United States (the United Nations is popular in most other countries) is due to its global standard-setting, which often clashes with corporate power.

States (the United Nations is popular in most other countries) is due to its global standard-setting, which often clashes with corporate power. Surveys by Globescan (www.globescan.org) regularly find public opinion in the sixty countries they survey favorable toward good corporate citizens as well as the United Nations (see also www.worldpublicopinion.org).

Linda Crompton (chapter 9) obtained a bird's-eye view during her leadership of IRRC. "The issue of regaining trust in corporations is obviously on everybody's minds at the moment, because it's very clear the markets have not recovered yet from all the recent scandals. There is a concern amongst investors in general that there are going to be yet more revelations. Everyone's watching the papers to see Eliot Spitzer's latest target. There's only one answer in terms of getting investors' confidence back: people will not have confidence unless they can believe what they're being told. If they have the impression that information is being kept from them and decisions being made that are not in their best interest—then investors will not return to the marketplace. My own theory—and a lot of other people feel the same way—is that this accounts for the shift of investment into property. People don't have a lot of other places to put their money. They will not return to the marketplace until they feel they can rely on what they're being told. Ultimately, the investor has a tremendous amount of power. They just haven't realized it and that's why I think you've seen the rise in shareholder activism. Now large institutional investors like CALPERS and others have started to flex their muscles. They're realizing they can. When you go to a company and you represent an enormous block of shares and you say, we don't like this, we want to see a change, companies listen."

Of the some hundred million Americans who are invested in the stock market, the half that are involuntary investors are unaware of what's happening with their money—or their rights regarding it. As mentioned in chapter 2, they have suffered heavy losses, through corporate bankruptcies and the meltdowns of Enron, WorldCom, Tyco, and many other companies. The ratings agency Standard & Poor's calculated that the companies in its 500-stock Index owe $442 billion more in retirement benefits than they have put aside. Of this amount, almost $300 billion is not reflected in their balance sheets (*Business Week*, Jan. 30, 2006).

Along with expensing stock options (i.e., putting these liabilities on balance sheets), the Securities and Exchange Commission is now encouraging the Financial Accounting Standards Board (FASB) to require that these unfunded pension liabilities also be on the balance sheet. This will be a big hit to the profits of major corporations—General Motors, Boeing, Verizon, IBM, AT&T, General Electric, United Technologies, Maytag, Goodyear, Ford, Navistar, Rockwell, Dara, and Visteon. Many, including Professor Don A. Moore at Carnegie Mellon University, say that the Sarbanes-Oxley law passed after the Enron meltdown, did not go far enough. In an editorial in *Business Week* (Apr. 17, 2006), he calls for more stringent oversight. Corporations are fighting back, often hiring hardball public relations firms to smear whistle blowers, such as Enron's Sherron Watkins and broadcaster Bill Moyers, as documented by *Business Week*'s story on Eric Denzenhall, "The Pit Bull of Public Relations" (Apr. 17, 2006). Other corporations decide to go private, delisting from stock exchanges or teaming up with hedge funds and private equity firms where disclosure is less rigorous. Hedge funds now account for over $1.1 trillion and have become shareholder activists of a different stripe: forcing companies through various raiding strategies to take on more debt and hype their short-term profits and stock prices. Yet this hedge fund game is already overcrowded and returns are sagging (*Business Week*, Jan. 30, 2006). Indeed the rush to financial and technological globalization may be reaching its limits. New vulnerabilities abound from financial crises, resource wars over oil, and dwindling water supplies to the ever-longer, more fragile global supply chains on which global corporations now rely, described in a special report in *The Economist* (June 17, 2006).

Phil Angelides (chapter 9) of the California Employee Pension Fund—CALPERS—explains the challenges funds like his must face. In the fourteen states that allow their employee beneficiaries to choose socially responsible investments, these pension funds may be less at risk. Phil is also California's state treasurer and has thought a lot about how best to protect public employees' retirements. "CALPERS has been involved because there's nothing more important to our overall portfolio than financial markets or corporate boardrooms that are viewed and are really transparent and honest and work well. After the market crash of 1929, a whole generation of Americans never invested in the stock market and our economy paid a dear price for that. It wasn't until 1954 that the Dow Jones Industrials exceeded its 1929 peak because people would rather put money in their mattress than in the stock market, which they saw as a rigged game for the insiders. So at our pension funds, we've been very strong on the reform movement—trying to open up the corporate boardroom, trying to have more transparent decisions, trying to make sure investment banks don't have conflicts of interest, and trying to make sure mutual funds really do serve the interests of customers, because our financial markets and our free enterprise system are real treasures. Our pension fund returns are tied, more than anything else, to the overall health of our economy. Therefore, we take a broader view now of prudence and responsibility."

Why do so many SRI asset managers' portfolios outperformed traditional ones? Socially responsible companies simply exhibit better overall management.

The SRI Advantage (2003) edited by Peter Camajo, founder of Progressive Asset Management, now in fourteen states, assembled papers by leading practitioners of SRI and reviewed how and why so many SRI asset managers' portfolios had outperformed traditional ones. Their conclusion: socially responsible companies simply exhibited better management overall. Investors seem to agree. According to mutual fund rating agency Lipper, assets in socially responsible mutual funds have jumped 156 percent over the past five years to nearly $32 billion while the fund industry as a whole has grown just 22 percent. The first U.S. investment fund to use social screens was the Pioneer Fund. Established during prohibition in 1929, it avoided investing in tobacco and alcohol companies.

SRI has helped drive shareholder advocacy and community investment and now it's gaining traction in Asia through the screening of Asian companies by ASRIA, founded by British SRI leader, Tessa Tennant (chapter 7). ASRIA (www.asria.org) audits their social, ethical, and environmental performance for many global banks, mutual funds, pension plans, and insurance companies. Due to the size and scope of green innovation and production worldwide, SRI venture capital funds are nurturing sustainable start-up companies in Asia, Latin America, Europe, as well as Canada and the United States. Nicholas Parker of Cleantech also coauthored with Diana Propper de Callejon *Cleantech Venture Investment: Patterns and Performance* (2005). "Cleantech investing and venturing is when people put risk capital to work in promising young companies which have the potential for highly disruptive solutions to old ways of doing things, for example, solar energy, clean water, things that we need and want. The challenge is to make them affordable and reliable. Venture investors are helping to make this happen." Nick explains, "The wonderful thing about Cleantech is that *by its very nature* it's responsible. If you're helping to commercialize technology that's disrupting old, dirtier ways of doing things, you're going to make world a better place! Plus, with venture capital it's about creating high quality jobs, so you're creating wealth, creating jobs, creating solutions to social problems, and eliminating environmental damage. This is why responsible investment in the Cleantech Venture world is so exciting! For the first time, we've got a comprehensive study that shows investors can make money from Cleantech venturing over a long period of time and in many countries in the world."

Deborah Sawyer (chapter 7) explains the financial obstacles in starting her firm, Environmental Design International, "The environmental or the hazardous waste business is about twenty-five years old. That's about how long I've been in the business. The moment I finished graduate school, I was hired literally the day I graduated by the Ohio Environmental Protection agency. Here I am, a kid right out of grad school, twenty some years old and getting to do all this high-powered stuff. It felt pretty heady! But I also had a degree in Political Science and a Masters degree in Biology, so I was a person who understood both sides of the law and the science. So it was a great opportunity. So that's how I got into the environmental field. I was lucky to have had some really good teachers. My mother has a math brain. I have a math brain. And I wouldn't let people tell me I wasn't good at things. Or when people do, that's actually what makes me work all that much harder. The challenge of the early days was financing. That is no longer really an issue for us. It's kind of funny, before, I had to beg for $50,000, today, when it's time to refinance, (we borrow about $4 million a year) it's all different. It took about five minutes to sign $4 million worth of loan documents. I used to beg and plead for $50,000, but now when it's loan renewal time, I have people wining and dining me on the fifty-seventh floor." It's good news for society that all these companies in the emerging sustainability sectors of the global economy are, at last, getting the financing they need to grow. Many will eventually provide many of the jobs of the future.

Mainstream financial media have finally woken up to the growing power of socially responsible investing (SRI), after a quarter-century of disdain and neglect. *Fortune*, (Feb. 7, 2005) which admits the growth of SRI, asked if green funds are true to their colors? noting they have different standards and definitions of social responsibility. Well, that's as it should be—since US investors have diverse goals and values. Some traditional economists and Wall Streeters continue to deny that such companies outperform those that focus on profit-maximization. As we have reported, this is despite decades of evidence showing that the new triple bottom line: people, profits, and planet, often leads to better financial performance. Why is this so? It's common sense really. Such companies turn out to simply have better management. Focusing on those three bottom lines enables management to develop broader vision and see further down the road. Short-term profit taking often leads to cutting corners—passing the company's costs on to others and the environment. Investors now look very carefully for such social and environmental liabilities overhanging balance sheets before they buy a company's stock. As we have seen, bad news travels fast in 24/7 markets and can quickly break a stock price or a precious brand name.

Most economics textbooks are simply decades out of date. They overlook the evolution of technology, markets, and societies—since steam engines in Britain three centuries ago. Economists still use static models of our dynamic evolving economies, as if they still were in general equilibrium, as if markets alone can manage these global changes. Economics is a noble profession, like law, medicine, and engineering, but it was never a science. Meanwhile, our twenty-first century marketplace keeps evolving as I have shown. I have described elsewhere, the new research by scientists from many other disciplines—mathematics, physics, biology, brain science, and ecology—has invalidated most of the core principles of traditional economic theory. The core principle most in error is the assumption about human nature, that individuals maximize their self-interest in competition with others. This bleak view of rational behavior also implies that cooperation sharing, unpaid caring, volunteering, nurturing children are by definition irrational. No wonder our societies look the way they do! Psychologist and futurist David Loye's *Darwin's Lost Theory of Love* (2004) finds that the great explorer and biologist was widely misinterpreted by Victorian elites in Britain who seized on the phrase, "the survival of the fittest" actually coined by Herbert Spencer, not Darwin, to justify their own privilege. Social Darwinism became fashionable while Adam Smith reinforced this idea of competition—albeit guided benevolently by an invisible hand—in his *Wealth of Nations* (1776). Such core assumptions underlying economic theory are also challenged from many directions and summed up by Robert Nadeau (chapter 1 in his *The Wealth of Nature* (2004)). Finally, in its issue of December 24, 2005, the London-based *The Economist* confessed that it was Herbert Spencer, one of their early contributors "who invented that poisoned phrase, 'the survival of the fittest'"—and admitted that cooperation was just as important as competition.

As David Loye points out, Darwin only used Spencer's phrase and mentioned competition less than ten times in his *The Descent of Man* and *The Origin of Species*, while referring to human cooperation hundreds of times. Darwin actually believed that human survival was largely due to our genius for cooperation. Brain research on mirror cells now confirms the role of empathy, generated by these mirror cells in human brains in the evolution of communication, cooperation, and culture (*Scientific American Mind*, Apr.–May 2006). The discovery in 1996 of mirror cells and their role is called a breakthrough that may turn out to be as big as the discovery of DNA in 1953. Indeed, as mentioned in chapter 1, in the historical evidence of human evolution we must conclude that cooperation was primary. Yet, we find our public and private decisions still dominated by that erroneous nineteenth-century economics.

The Age of Light

Emerging Lightwave Technologies (Photonics)

Fiber optics, Optical scanners, Lasers, Holography

Solar technologies

Optical computers, Multiprocessor, parallel computers and neural net computers, Imaging technologies

Biotechnologies

Gene machine, DNA sequencers, Tagging and tracking chemicals and genes, Nano technologies

Photons (sunlight) falling on the Earth supply enough energy in 10 minutes to put our entire six billion population in orbit!

© Henderson, 1991.

In reality, we know that humans enjoy sharing and cooperating as much as competing. Yet, competition is rewarded and reinforced by that malfunctioning economic source code—while unselfish, unpaid sharing, caring, and cooperating are punished by being ignored and uncounted in GNP/GDP indexes. Thus, feminist economists led by Marilyn Waring in her *If Women Counted* (1987) uncovered the truth: economics has been patriarchal to its core! As David Loye points out, Charles Darwin also viewed humans as capable of altruism and predicted the evolution of moral sentiments as part of our species' survival and evolution—a subject Adam Smith himself studied in his *Theory of Moral Sentiments*. If this is true, we can take heart that the evolution of markets toward more ethical behavior is simply a part of our evolution toward greater awareness and responsibility as a species. We see signs of this in such grassroots organizing worldwide to produce the Earth Charter (www.earthcharter.org), the U.N. Millennium Development Goals, and the Global Marshall

Plan for a clean, green global development (www.globalmarshall plan.org).

In the hearts of all the people you have met in these pages, a shared vision burns brightly. They see beyond the bloodshed and chaos of humanity's past mistakes to a future arising out of the deeper lessons we have learned. Technologically, and biologically, we became the most successful species on this planet—arising from the African continent to populate almost every land on the face of the Earth. Our consumption of 40 percent of all the planet's primary production of photosynthesis is leading to today's second greatest period of extinction of other life forms. With our enlarged forebrains and opposing thumbs, our technologies have enabled us to create undreamed of material well-being for one-third of our six-billion plus human family, explore the moon, and set our sights on Mars, and other planets.

Our new challenge is to face ourselves, our ancient fears of scarcity, the other and our own mortality, which drove competition, our territorial conflicts, injusticies, fratricidal wars, and genocide. Today, our increasingly crowded, polluted planet is mirroring back to humans the unsustainability of our fossil-fueled industrial technologies and lifestyles. The Earth's message was heeded by millions and in the late twentieth century the great transition to a new information based solar age began. Today's growing green economies everywhere embody deeper human learning about the functioning of our planetary life support system—the expansion of our time horizons and spatial awareness. This is leading to a rapid reintegration of human knowledge beyond the brilliant achievements of Cartesian, reductionist knowing. We are reconnecting the dots—learning the first law of ecology: Everything is connected to everything else. This post-Cartesian worldview is now driving the rapid growth of more sustainable forms of human development, whole-systems thinking, and integrated approaches to private and public decision-making. We have learned that narrow, fragmented, short-term thinking is now too costly and is unsustainable. When seen in a planetary context, we understand that all our self-interests are *identical*! Morality has become pragmatic.

Acknowledgments

I am indebted to all the members of the Research Advisory Board for *Ethical Markets* TV series and my Advisory Board for the Calvert-Henderson Quality of Life Indicators. All are listed on pages 234–236. Most of them have also been friends and mentors for many years. My life partner and spouse, Alan F. Kay, the Internet pioneer, has also partnered with me in launching several start-up companies, including Ethical Markets Media, LLC. Alan's experience as a mathematician, in business, technology, and opinion research, expanded my horizons. His loving understanding and emotional support have been a constant blessing.

Mary Bahr is the most wonderful editor with whom I have ever worked, and her enthusiasm for this book energized me enormously. I thank Margo Baldwin, John Barstow, Collette Leonard, Marcy Brant, and all at Chelsea Green for their dedication to envisioning a cleaner, greener, saner global economy and society.

My gratitude to Ellyne Lonergan, President, Glass Onion Productions, who coproduced the final *Ethical Markets* TV series with me in a deeply enjoyable collaboration. I salute Maureen Hart for so ably managing the Web site www.Calvert-Henderson.com for the Quality of Life Indicators. My thanks to Jan Crawford, my Executive Assistant, for managing the manuscript and the complex process of cataloging and organizing the illustrations for the book, as well as for managing the Web site www.EthicalMarkets.com. Laury Saligman applied her invaluable skills in marketing the TV series to helping assemble the pictures of all our eighty experts and leaders who appeared in the series and in this book. Laury was assisted by Jeremy Michael Cohen, who contributed his time voluntarily during a busy work and study schedule.

The host of our TV series, Simran Sethi, has a special place in my heart. Simran, who wrote many of the scripts for the series, earned her MBA in sustainable business. Our friend and writer of the Foreword, Hunter Lovins, was Simran's thesis advisor. Simran is so expert and so well-versed in my books and so many others that I expect she will spearhead much new thinking and action toward creating more ethical markets. I am grateful for her dedication and deep commitment to a brighter future for all on planet Earth.

Ethical Markets Research Advisory Board

Socially Responsible Investing

Joan Bavaria, President, Director & Cofounder, Trillium Asset Management Corporation, Boston, MA
Hal Brill, President, Natural Investment Services, Inc., Paonia, CO
Susan Davis, President, Capital Missions Co., Elkhorn, WI
Dr. Judy Henderson, Chair, Global Reporting Initiative, Canberra, Australia
Barbara Krumsiek, President & CEO, The Calvert Group, Bethesda, MD
Marcello Palazzi, President, Progressio Foundation, Netherlands
Robert Rubenstein, CEO & founder, Brooklyn Bridge and TBLI Group, Amsterdam, Netherlands
Steve Schueth, President, First Affirmative Financial Network, Boulder, CO
Timothy Smith, Senior Vice President, Walden Asset Mgmt; President, Social Investment Forum, Boston, MA
Rodger Spiller, Money Matters Inc., Auckland, New Zealand
Tessa Tennant, Director, ASRIA, Hong Kong
Eva Willmann de Donlea, President, Conscious Investments, Sydney, Australia

Corporate and Government Accountability Research

Susan Aaronson, Senior Fellow, Nat. Policy Assoc., Washington, DC
Verna Allee, author *Increasing Prosperity through Value Networks*, Martinez, CA
Gary Brouse, Director, Interfaith Ctr. on Corp. Responsibility, New York, NY
Kim Cranston, President, Institute for Organizational Evolution, CA
Linda Crompton, Sr. Advisor, Investor Responsibility Research Center, Washington, DC
Susan Davis, Advisor to the Director-General, ILO, Geneva, Switzerland; Chair, Grameen Foundation, United States
Ed Mayo, Chief Executive, National Consumer Council, London, United Kingdom
Cliff Feigenbaum, Editor, *Green Money Journal*, Santa Fe, NM
Prof. Michael Hoffman, Executive Director, Center for Business Ethics, Bentley College, Waltham, MA
Susan Kavanaugh, Managing Editor, *Federal Ethics Report*, Washington, DC
Matthew Kiernan, Managing Director, Innovest Inc., Toronto, Canada
Mindy Lubber, Executive Director, CERES, Boston, MA
Nell Minow, Cofounder, The Corporate Library, Washington, DC
Robert A.G. Monks, author The New Global Investors, Cape Elizabeth, ME
Darin Rovere, Ctr. For Innovation in Corp. Responsibility, Ottawa, Canada
Professor S. Prakash Sethi, President, International Center for Corporate Accountability, Inc., Baruch College/CUNY, New York, NY
Deirdre Taylor, Founder, *Spirituality & Health* magazine, Westport, CT
Alice Tepper Marlin, President, Social Accountability International, NY
Shann Turnbull, PhD, Principal, International Institute for Self-governance, Sydney, Australia
Toby Webb, Editor, *Ethical Corporation* magazine, London, United Kingdom

Sustainability/Renewable Resource Sectors

Ralph Abraham, PhD, Prof. of Mathematics, Univ. of California, Santa Cruz, CA
Christine Austin, Teal Onstad Productions, Leverett, MA
Tariq Banuri, Stockholm Environment Institute, Pathumthani, Thailand
Jacqueline Cambata, President, Triad Co. Ltd., Vienna, VA
Craig Cox, Editor, *UTNE READER*, Minneapolis, MN
Peter Davies, OBE, Deputy Chief Executive, Business in the Community, London, United Kingdom
Prof. Michael Dorsey, Dept. of Environmental Studies, Dartmouth College, Hanover, NH
Riane Eisler, author, *The Chalice and The Blade*; President, Center for Partnership Studies, Pacific Grove, CA
Duane Elgin, author, *Promise Ahead*; *Voluntary Simplicity*; *Awakening Earth*; Cofounder of Our Media Voice; San Anselmo, CA
Carl Frankel, Sr. Editor, Green@Work, author, *In Earth's Company*, Kingston, NY
Lori Grace, MA (Psych), Sunrise Center, Corte Madera, CA
Alisa Gravitz, President, Co-op America, Washington, DC
Denis Hayes, President, Bullitt Foundation, Seattle, WA
Tachi Kiuchi, Chairman, Future 500, Tokyo, Japan
Francis Koster, Chief, Innovation Services, Nemours Foundation, FL
Walter Link, Chairman, The Global Academy, Mill Valley, CA
Richard Lippin, MD, President, The Lippin Group, PA
Elsie Maio, President, Maio & Co. (Mgmt. Consultants in Ethical Branding), NY

Michael Marvin, Past-president, Business Council for Sustainable Energy, Washington, DC
Paul E. Metz, PhD, President, European Business Council for a sustainable Energy Future, Brussels, European Union
Prof. Bedrich Moldan, Director, Environment Center, Charles U., Czech Republic
Jim Motavalli, Editor, *The Environmental* magazine, CT
Prof. Robert Nadeau, author, *The Wealth of Nature*, Geoge Mason Univ., VA
F. Byron Nahser, Provost, Presidio World College, San Francisco, CA
Claudine Schneider, Sr. VP, Econergy Int'l Corp., Boulder, CO
Rena Shulsky, Founder, GREEN SEAL, Inc; President, Shire Realty, NY
Carol Spalding, PhD, President, Open Campus, Florida Community College of Jacksonville, FL
Radames Soto, President, Kinina Ventures, Miami, FL
Stephen Viederman, Past President, Jessie Smith Noyes Foundation, NY
Richard Wilson, UK Business Council for Sustainable Energy, London, United Kingdom
Jane Zhang, CEO, Sino-American Dev. Corp, NY & Shanghai, China
Zhouying Jin, Professor, Innovation & Strategies, Chinese Academy of Social Sciences, China

Social Venture Capital

Mark Donohue, Expansion Capital Partners, San Francisco, CA
Elizabeth Harris, UNC Partners, Inc., Boston, MA
Tony Lent, EA Capital, New York, NY
Nicholas Parker, Cofounder & Chairman, CLEANTECH Venture Network LLC, Toronto, Canada
D. Wayne Silby, Esq., President, Calvert Ventures, LLC & Founder, Calvert Social Investment Funds, Washington, DC
Stuart Valentine, President, Iowa Progressive Asset Management, Fairfield, IA
Woody Tasch, Chair, Investors Circle, MA
Steve Waddell, Executive Director, GAN-NET, Boston, MA
Eckart Wintzen, President, Ex'tent Management & Investment Co., Netherlands

Globalization/North-South Equity, Justice

Rosa Alegria, President, Perspektiva Co., Sao Paulo, Brazil
Prof. Walden Bello, Codirector, Focus on the Global South, Chulalongkorn University, Bangkok, Thailand
Frank Bracho, Former Ambassador of Venezuela to India, author, *Hacia Una Forma de Medir el Desarrollo*, Caracas, Venezuela
Rinaldo Brutoco, President, World Business Academy, Santa Barbara, CA
Fritjof Capra, author *The Web of Life* and *The Hidden Connections*, Berkeley, CA
Christina Carvalho Pinto, President, Full Jazz Advertising Agency, Sao Paulo, Brazil
Thais Corral, Director General, REDEH, Rio de Janeiro, Brazil
Prof. Michael Hopkins, PhD, MHCI Ltd., author *The Planetary Bargain: CSR Matters*, London, United Kingdom & Geneva, Switzerland
Paul Freundlich, President, The Fair Trade Foundation; Board member, CERES, GRI
Devaki Jain, author, *Women Development and the UN*, Bangalore, India
Rushworth M. Kidder, President, Institute for Global Ethics, Camden, ME
Ashok Khosla, CEO, Development Alternatives, New Delhi, India
Annabel McGoldrick, Peace Journalism, Oxford, United Kingdom
Oscar Motomura, CEO, Amana-Key Desenvolvimento & Educacao Ltda, Saõ Paulo, Brazil
Jane Nelson, Harvard University and Prince of Wales Business Leaders Forum, London, United Kingdom
Peter Orne, Editor, *WorldPaper*, Boston, MA
David Roodman, Center for Global Development, Washington, DC
Ziauddin Sardar, Editor, Futures Elsevier, Ltd, London, United Kingdom
Elisabet Sahtouris, Evolution biologist, author, Santa Barbara, CA
Crocker Snow, Founder, Money Matters Inst., Boston, MA
Miklos Sukosd, PhD, Dept. Political Science, Central European University, Budapest, Hungary
Wouter Van Dieren, President, Institute for Environment & Systems Analysis, Amsterdam, Netherlands
Judith Woodard, Christopher Street Financial Co., New York, NY
Norio Yamamoto, PhD, Exec. Vice President, Global Infrastructure Fund Research Foundation, Tokyo, Japan
Susana Oseguera Yturbide, Communications Internacionales, Mexico

Local/Community Investing

Rebecca Adamson, President, First Nations Development Inst., United States
Rosalind Copisarow, Founder, Fundusz Mikro, Warsaw, Poland and London, United Kingdom
Lynne Franks, CEO, Seed Fusion, London, United Kingdom
Sharon Hadary, Exec. Dir., Center for Women's Business Research, Washington, DC
Robert MacGregor, Pres Emeritus, Center for Ethical Business Cultures, Minn., MN
Terry Mollner, Trustee, Calvert Social Investment Fund, United States
Pam Solo, President, Institute for Civil Society, Boston, MA
Michaela Walsh, Founding President, Women's World Banking, NY
Julie Weeks, President, Womenable, Empire, MI

Calvert-Henderson Quality of Life Indicators Advisory Board

Selected Bibliography

Global Transition

Barnes, Peter. *Capitalism 3.0*, Berrett Koehler, San Francisco, CA, 2006.
Boulding, Elise. *Building a Global Civic Culture*. New York: Columbia University Press, 1988.
———. *Cultures of Peace*. New York: Syracuse University Press, 2000.
Bracho, Frank. *Petroleo Y Globalizacion: Salvacion or Perdicion?* Caracas: Vadell, 1998.
Corcoran, Peter Blaze, ed. *The Earth Charter in Action*. Amsterdam, Netherlands: Kit Publishers, 2005.
Harman, Willis. *Global Mind Change*. Foreword by Hazel Henderson. San Francisco: Berrett-Koehler, 1998.
Henderson, Hazel. *Politics of the Solar Age*. New York: Doubleday, 1981; New York: Knowledge Systems/TOES Books, 1988.
———. *Paradigms in Progress*. San Francisco: Berrett-Koehler, 1991, 1995.
———. *Building a Win-Win World*. San Francisco: Berrett-Koehler Publishers, 1996.
———. *Beyond Globalization*. Bloomfield, CT: Kumarian Press, Inc., 1999.
Henderson, Hazel, Harlan Cleveland, and Inge Kaul, eds. "United Nations: Policy and Financing Alternatives," 1st ed., *Futures*, London: Elsevier Scientific, 1995; U.S. ed., Washington, DC: Global Commission to Fund the U.N., 1996.
Henderson, Hazel and Daisaku Ikeda. *Planetary Citizenship*. Santa Monica, CA: Middleway Press, 2004.
Houston, Jean. *Jump Time*. Putnam, NY: Jeremy Tarcher, 2000.
Independent Commission on Quality of Life. President, Pinta Silgo, Maria deLourdes. *Caring for the Future*. Oxford: Oxford University Press, 1996.
Jain, Devaki. *Women, Development and the UN*. Bloomington, IN: Indiana University Press, 2005.
Kaul, Inge and Pedro Conceicao, eds. *The New Public Finance*, Oxford University Press, 2006.
Landes, David S. *The Wealth and Poverty of Nations*. New York: W. W. Norton, 1998.
Marx Hubbard, Barbara. *Conscious Evolution*. Novato, CA: New World Library, 1998.
Miller, John H. ed. *Curing World Poverty*. St. Louis, MO: Social Justice Review, 1994.
Prestowitz, Clyde. *Three Billion New Capitalists*. New York: Perseus Books, 2005.
Rifkin, Jeremy. *The End of Work*. Los Angeles, CA, Jeremy Tarcher, 1995.
Sachs, Jeffrey. *The End of Poverty*. New York: Penguin Press, 2005.
Smith, Stephen C. *Ending Global Poverty*. New York: Palgrave, Macmillan, 2005.
The Group of Lisbon. *The Limits to Competition*. Cambridge, MA: MIT Press, 1995.
United Nations. *Agenda 21*. New York: United Nations, 1992.
United Nations Development Program. *Human Development Reports*. New York: United Nations Development Program, 1990–2006.
Von Weizsacker, Ernst Ulrich, Oran Young, and Matthias Finger, eds. *Limits to Privatization*. London: Earthscan, 2005.
Williamson, Marianne, ed. *Imagine*. Emmaus, PA: Rodale, 2000.
World Commission on Environment & Development, Chair Brundtland, Gro Harlem. *Our Common Future*. Oxford: Oxford Press, 1987.

New Science

Allee, Verna. *The Future of Knowledge*. Amsterdam, NL: Elsevier Science, 2003.

Barabasi, Albert-Laszlo. *Linked*. Cambridge, MA: Perseus, 2002.

Benyus, Janine M. *Bimimicry*. New York: Wm Morrow, 1997.

Bornstein, David. *How to Change the World*. Oxford: Oxford University Press, 2004.

Brown, Lester R. *Eco-Economy*. New York: W.W. Norton, 2001.

Capra, Fritjof. *The Web of Life*. New York: Anchor Doubleday, 1996.

———. *The Hidden Connections*. New York: Anchor Doubleday, 2002.

Daly, Herman F. and John B. Cobb, Jr. *For the Common Good*. Boston: Beacon Press, 1989.

Dowling, Keith, Jurgen de Wispelaere, and Stuart White, eds. *The Ethics of Stakeholding*. United Kingdom: Anthony Rowe, 2003.

Fullbrook, Edward, ed. *A Guide to What's Wrong with Economics*. London: Anthem Press, 2004.

Hawken, Paul, Amory Lovins, and Hunter Lovins. *Natural Capitalism*. New York: Little Brown, 1999.

Henderson, Hazel and Calvert Group. *Calvert-Henderson Quality of Life Indicators*. Bethesda, MD: Calvert Group, 2000. (updated at www.calvert-henderson.com)

Keen, Steven. *Debunking Economics*. Australia: Pluto Press, 2005.

Kiuchi, Tachi and Bill Shireman. *What We Learned in the Rainforest*. San Fransisco: Berrett Koehler, 2002.

Kurland, Norman, Dawn K. Brohaun, and Michael D. Greaney. *Capital Homesteading*. Washington, DC: Economic Justice Media, 2004.

Loye, David. *Darwin's Lost Theory of Love*. New York: to Excel, 2000.

McDonough, William, and Michael Braumgart. *Cradle to Cradle*. NY: North Point Press, 2002.

Nadeau, Robert. *The Wealth of Nature*. New York: Columbia Univ. Press, 2003.

Nadeau, Robert and Menas Kafatos. *The Non-Local Universe*. Oxford: Oxford University Press, 1999.

Sahtouris, Elisabet. *Earthdance: Living Systems in Evolution*. Santa Barbara, CA: Metalog, 1995.

Schumacher, E.F. *Small is Beautiful*. Harper, 1975, New York: Blond & Briggs, London, 1973.

Wackernagel, Matthis, and William Rees. *Our Ecological Footprint*. Vancouver, BC: New Society Publishers, 1996.

Ethical Business

Abrams, John. *The Company We Keep*. White River Junction, VT: Chelsea Green, 2005.

Albion, Mark. *True To Yourself*. San Fransisco: Berrett-Koehler, 2006.

Brill, Hall, Jack A. Brill, and Cliff Feigenbaum. *Investing With Your Values*. Vancouver, BC: New Society Publishers, New Edition, 2000.

Camajo, Peter, ed. *The SRI Advantage*. Gabriola Island, BC, Canada: New Society Publishers, 2002.

Cohen, Ben, and Mal Warwick. *Values-Driven Business*. San Francisco: Berrett-Koehler, 2006.

Domini, Amy. *Socially Responsible Investing*. Boston, MA: Dearborn Trade/Kaplan, 2001.

Franks, Lynne. *The Seed Handbook: The Feminine Way to Create Business.* New York: Tarcher Putnam, 2000.
Hart, Stuart L. *Capitalism at the Crossroads.* Philadelphia, PA: Wharton School Publishing, 2005.
Hollender, Jeffrey, and Stephen Fenichell. *What Matters Most.* New York: Basic Books, 2004.
Huff, Priscilla Y. *Her Venture.com.* Roseville, CA: Prima, 2000.
Kelso, Louis O., and Patricia Hetter. *Two-Factor Theory: How to Turn Eighty Million Workers into Capitalist on Borrowed Money.* New York: Vintage, 1967.
———. *Democracy and Economic Power.* New York: Ballinger, 1986.
Maloney, Julie, and Renee Moorefield. *Driven By Wellth.* Boulder, CO: Wellth Productions, 2004.
Nelson, Jane, and Ira J. Jackson. *Profits with Principles.* New York: Doubleday, 2004.
Parker, Thornton. Foreword by Hazel Henderson. *What If the Boomers Can't Retire?* San Francisco: Berrett Koehler, 2000.
Pralahad, C. K. *The Fortune at the Bottom of the Pyramid.* Philadelphia, PA: Wharton School Publishing, 2002.
Secretan, Lance H. K. *Reclaiming Higher Ground.* New York: McGraw Hill, 1997.
———. *Inspire!* New York: John Wiley, 2004.
Seger, Linda. *Web Thinking.* HI: Inner Ocean, 2002.
Shuman, Michael H. *The Small-Mart Revolution.* San Francisco: Berrett Koehler, 2006.
Tapscott, Don, and David Ticoll. *The Naked Corporation.* New York: Free Press, 2003.
Co-op America. *The Green Pages.* Washington, DC. Published annually.

Redefining Success

Burns, Scott. *The Household Economy.* New York: Doubleday, 1979.
Benkler, Yochai. *The Wealth of Networks.* Yale University Press, New Haven, CT, 2006.
Cahn, Edgar S. *No More Throwaway People.* Washington, DC: Essential Books, 2004.
———. *How to Manual: The Time Dollar.* Washington, DC: Time Dollar Institute, 2005.
Eisler, Riane. *The Real Wealth of Nations.* San Francisco: Berrett-Koehler, 2007.
———. *The Power of Partnership.* Novato, CA: New World Library, 2002.
Hyde, Lewis. *The Gift.* New York: Vintage Books, 1979.
Layard, Richard. *Happiness: Lessons from a New Science.* London: Penguin Group, 2005.
Mitchell, Ralph, and Neil Shafer. *Depression Scrip of the USA, Canada & Mexico.* Iola, WI: Krause Publications, 1984.
Shiva, Vandana. *Staying Alive.* London: Zed Books, 1989.
Twist, Lynne. *The Soul of Money.* New York: W. W. Norton, 2003.
Toffler, Alvin and Heidi. *Revolutionary Wealth.* Knopf, New York, 2006.
Vaughan, Genevieve. *For-Giving.* Austin, TX: Plain View Press, 1997.
Waring, Marilyn. *If Women Counted.* New York: Harper & Row, 1988.
Zarlenga, Stephen. *The Lost Science of Money.* Valatie, NY: American Monetary Institute, 2002.

New Politics

Adams, Michael. *American Backlash.* Toronto, Canada: Viking Press, 2005.
Armstrong, Jerome, and Markos Moulitsas Zuniga. *Crashing the Gate.* White River Junction, VT: Chelsea Green, 2006.

Bardon, Doris and Laurie Murray. *Creative Leadership for Community Problem Solving.* Gainsville, FL: Institute for Creative Leadership, 2006.

De Soto, Hernando. *The Other Path.* New York: W. W. Norton, 2003.

Emanuel, Rahm and Bruce Reed. *The Plan: Big Ideas for America.* New York, NY: Public Affairs, 2006.

Hallsmith, Gwendolyn. *The Key to Sustainable Cities.* Vancouver, BC: New Society Publishers, 2003.

Kay, Alan F. *Locating Concensus for Democracy.* St. Augustine, FL: Americans Talk Issue Foundation, 1998.

———. *Spot the Spin.* Victoria: Trafford, 2004.

Kochkin, Alex S., and Patricia VanCamp. *A New America.* Point Reyes, CA: Fund for Global Awakening Survey Research, 2000–2005.

Kohn, Alfie. *No Contest: The Case Against Competition.* New York: Houghton Mifflin, 1986.

Lake, Celinda, and Kellyanne Conway. *What Women Really Want.* New York: Free Press, 2005.

Laszlo, Ervin. *You Can Change the World.* Place: Club of Budapest, 2003.

LeRoy, Greg. *The Great American Jobs Scam.* San Francisco: Berrett-Koehler, 2005.

Mills, Stephanie. *Epicurean Simplicity.* Washington, DC: Island Press, 2001.

Ray, Paul H., and Sherry R. Anderson. *The Cultural Creatives.* New York: Harmony Books, 2000.

Key Web Sites

- AccountAbility (accountability.org.uk)
  Performs social audits on companies globally
- Ashoka (Ashoka.org)
  Highlights and promotes social entrepreneurism globally
- Association for Sustainable & Responsible Investment in Asia (ASRIA.org)
  Researches and rates Asian companies' social performance
- Bainbridge Graduate Institute (bgiedu.org)
  Offers an MBA in sustainability management
- Business Alliance for Local Living Economies (livingeconomies.org)
  Association to build local economies & support local communities
- *Business Ethics* (business-ethics.com)
  Leading industry magazine on corporate responsibility
- Business for Social Responsibility (BSR.org)
  Big-business-oriented group on corporate responsibility issues
- Calvert-Henderson Quality of Life Indicators (calvert-henderson.com)
  Hazel Henderson's indicators developed with the Calvert Group that offer statistics on twelve aspects of wealth beyond GDP
- Center for Business as Agent of World Benefit (worldbenefit.cwru.edu)
  A new business school at Case Western University
- Center for Integrity in Science (integrityinsicence.org)
  Research on academic-corporate links
- Center for Media Transparency (mediatransparency.org)
  Research on how media are financed
- Center for a New American Dream (newdream.org)
  Key group redefining success and living sustainably
- Center for Public Integrity (cspi.org/integrity)
  Best research on corporate and US government accountability
- Centre for Science and Environment (cseindia.org)
  Environmental impacts of business within India
- CERES (ceres.org)
  Coalition of environmentally-concerned pension funds
- Clean Edge (cleanedge.com)
  Covers renewable energy, offering sound information for investors and the rest of us
- Cleantech Venture Network (cleantech.com)
  Key group of "green" venture investors
- Coop America (coopamerica.org)
  Info on socially-responsible investing, green businesses, and practical steps each of us can take to live more sustainably
- Corp Watch (corpwatch.org)
  Grassroots watchdogs holding corporations accountable globally
- CSR Newswire (csrwire.com)
  Newswire service on corporate social responsibility
- *Dollars and Sense* (dollarsandsense.org)
  Online version of the magazine on economic literacy and justice
- *Dwelling* (dwelling.com)
  Mainstream magazine on green building and design
- Earth Charter (earthcharter.org)
  Global grassroots principles of Earth Ethics
- Earth Policy Institute (earth-policy.org)
  Lester R. Brown's recommendations for steering toward sustainability
- *E-Magazine* (emagazine.com)
  Premier national environmental magazine
- Ethical Corporation (ethicalcorp.com)
  Comprehensive overview of global corporate responsibility
- *Ethical Markets* (ethicalmarkets.com)
  The first national portal and television series on corporate social responsibility, ethical investing
  and the "green" sectors on PBS affiliates throughout the United States
- Instituto Ethos (ethos.org.br)
  Premier group of socially-responsible businesses in Brazil
- Equal Access (equalaccess.org)
  NGO providing access to global communications for rural communities in the south
- Fair Economy (faireconomy.org)
  NGO campaigning and research for fairness in the United States
- Green Biz (greenbiz.com)
  News source on sustainable business
- *Green Economics* (greeneconomics.org.uk)
  Scholarly papers on sustainable economics
- Global Reporting Initiative (globalreporting.org)
  Promotes "triple-bottom line" accounting for annual corporate reports
- Global Transition Initiative (gti.org)
  Global network on the transition to sustainability

- Innovest (innovestgroup.com)
  Premier social and environmental firm auditing corporations globally
- Investor's Circle (investorscircle.net)
  Venture capital for social entrepreneurs and sustainable business
- Institute of Noetic Sciences (noetic.org)
  Membership group on shifting human consciousness toward planetary concerns
- Mothering (mothering.com)
  Action alerts that help create a better world for us and our kids
- Natural Capitalism Solutions (natcapsolutions.com)
  Promoting sustainability in business through the principles of Natural Capitalism led by Hunter Lovins
- Navdanya (navdanya.org)
  Vandana Shiva's site on food justice, seed sovereignty, the organics movement worldwide
- Net Impact (netimpact.org)
  Site for MBAs dedicated to leveraging business for social change
- *ODE* (odemagazine.com)
  Global news on efforts to create a better world for all
- Pesticide Action Network North America (panna.org)
  Resource for information on pesticides and pesticide alternatives
- Presidio World College (presidiomba.org)
  Offers a MBA in sustainable business
- Principles for Responsible Investment Group (unpri.org)
  Pension funds group representing over $4 trillion in assets in sixteen countries
- Responsible Shopper (responsibleshopper.org)
  Co-op America on how to avoid social and environmental impacts of corporations
- *Resurgence* (resurgence.org)
  Britain's key magazine on global sustainability
- SA International (sainternational.us)
  Premier watchdog on global labor standards
- Simplicity Forum (simpleliving.net/simplicityforum)
  Ideas on sustainable consumption
- Social Investment Forum (socialinvest.org)
  Socially responsible investing providers and national research
- SustainAbility (sustainability.com)
  Think tank consultancy on risk management, sustainable business practices
- Treehugger.com
  Popular blog for green product experiences
- UN Environment Program Finance Initiative (unepfi.org)
  The best source on global environmental finance initiatives
- UN Global Compact (unglobalcompact.org)
  Over two thousand corporations have signed on to their Ten Principles of Corporate Citizenship
- World Business Academy (www.worldbusiness.org)
  Network of responsible business leaders
- World Business Council for Sustainable Development (wbcsd.org)
  Business and sustainable development all over the world
- World Resources Institute (wri.org)
  Information on topics including biodiversity, food, climate change, etc.
- Worldwatch Institute (worldwatch.com)
  Global sustainability researchers

Note: Hazel Henderson's articles, papers, editorials are at: www.hazelhenderson.com

Photo Credits

Chapter 1

p. 2, Gross National Product Problems	© Hazel Henderson 1979
p. 3, Betsy Taylor, Center for a New American Dream	Center for a New American Dream
p. 5, Inge Kaul, UN Development Program	UN Development Program
p. 7, Mathis Wackernagel, Ecological Footprint Analysis	Ecological Footprint Network
p. 13, Vidette Bullock Mixon	United Methodist Church
p. 14, Nobel Prize Medal	Nobel Committee
p. 15, Professor Ralph Abraham	University of California
p. 16, Professor Robert Nadeau	George Mason University
p. 19, Ray Anderson with Simran Sethi	Ethical Markets Media, LLC
p. 20, Hewson Baltzell	Ethical Markets Media, LLC

Chapter 2

p. 25, Rich Ferlauto	AFSCME
p. 26, Alisa Gravitz	Co-op America
p. 29, Jane Nelson	Jane Nelson
p. 30, Georg Kell, Executive Head	UN Global Compact
p. 33, Mallen Baker	Business in the Community, UK
p. 34, Oded Grajew	Ethos Institute, Brazil
p. 38, Simran Sethi, Host	Ethical Markets Media, LLC
p. 38, Alex Counts	Grameen USA
p. 38, Alice Tepper-Marlin, President	Social Accountability International, NY
p. 40, Fruit Seller	Grameen USA

Chapter 3

p. 46, Scott Burns	Dallas Morning News
p. 47, Rebecca Adamson, First Nations Development Institute	Photographer: Doug Barber and The Calvert Group, Bethesda, MD
p. 48, Riane Eisler	Center for Partnership Studies
p. 49, Edgar Cahn, Founder	Time Dollar Institute
p. 52, Vandana Shiva	Research Foundation for Science, Technology and Ecology
p. 54, Total Productive System of an Industrial Society	© Hazel Henderson 1981
p. 56, Simran with Bob Meyer	Ethical Markets Media, LLC
p. 56, Hewson Baltzell	Ethical Markets Media, LLC

Chapter 4

p. 60, Adam Joseph Lewis Center for Environmental Studies at Oberlin College	Oberlin College, Oberlin, OH
p. 60, William McDonough, Architect	William McDonough & Partners
p. 61, Ford's Green Plant	William McDonough & Partners
p. 61, Hunter Lovins	Natural Capitalism Solutions
p. 63, Kathleen Hogan	US EPA
p. 63, EPA's Energy Star Label	US EPA
p. 64, NYC Habitat Home	Habitat for Humanity
p. 66, Leslie Hoffman on Earth Pledge Green Roof, NYC	Earth Pledge

p. 67, Earth Pledge's Green Roof, NYC	Earth Pledge
p. 72, Simran Sethi, Host	Ethical Markets Media, LLC
p. 72, George Terpilowski	Fairmont Hotels
p. 73, Fairmont Hotel, Hawaii	Fairmont Hotels

Chapter 5

p. 77, Judy Wicks and Friends	White Dog Café
p. 78, William Drayton	Ashoka
p. 80, Susan Davis	Grameen, USA
p. 81, Shari Berenbach	Calvert Foundation
p. 83, Ronni Goldfarb's friends in Afghanistan	Equal Access
p. 88, Simran with Hewson Baltzell and Jean Pogge	Ethical Markets Media, LLC

Chapter 6

p. 92, Fair Trade Certified Label	Transfair USA
p. 93, Paul Rice	Transfair USA
p. 95, Chris Mann	Guayaki Company
p. 95, Harvesting Mate	Guayaki Company
p. 96, Amber Chand	The Jerusalem Candle
p. 101, Principles of Sustainable World Trade	© Hazel Henderson 2002
p. 103, Simran Sethi	Ethical Markets Media, LLC
p. 103, Robert Stiller	Green Mountain Coffee Company

Chapter 7

p. 108, Deborah Sawyer	Environmental Design International
p. 109, Judy Wicks	White Dog Café
p. 114, Paul Ray	Paul Ray
p. 115, Nancy Barry	Womens World Banking
p. 116, Michaela Walsh	Womens World Banking
p. 117, Susan Davis	Grameen USA
p. 120, Alice Tepper Marlin, Simran Sethi, and Amy Hall	Ethical Markets Media, LLC

Chapter 8

p. 126, Geoffrey Ballard	Ballard Power Company
p. 127, Mindy Lubber	CERES
p. 131, Susan Davis	Capital Missions Company
p. 132, Amory Lovins	Rocky Mountain Institute
p. 133, Rocky Mountain Institute, Colorado	Rocky Mountain Institute
p. 135, Blue Energy Canada's Tidal Fence	Blue Energy, Vancouver, Canada
p. 135, Bob Freling	Solar Electric Light Fund
p. 139, Simran Sethi w/Mark Farber, Evergreen Solar Corp.	Ethical Markets Media, LLC

Chapter 9

p. 146, Linda Crompton	Investor Responsibility Research Center
p. 147, Gary Brouse	Interfaith Center for Corporate Responsibility
p. 150, Phil Angelides	CALPERS
p. 152, Corporations Stakeholders	© Hazel Henderson 1991
p. 155, Jennifer Barsky, Simran Sethi, and Roberta Karp	Ethical Markets Media, LLC

Chapter 10

p. 161, Jeremy Rifkin — Foundation for Economic Trends
p. 163, Patricia Kelso and Louis O. Kelso — Kelso Institute
p. 165, Bernie Glassman — Greyston Bakery
p. 169, Gary Erickson — Clif Bar Company
p. 171, Lynne Twist — Soul of Money Institute
p. 175, Simran Sethi, Paul Millman, and Paul Freundlich — Ethical Markets Media, LLC
p. 177, Employee-owners — Chroma Technology Corporation

Chapter 11

p. 181, Thomas Fricke — Forestrade, Inc.
p. 183, Gary Hirshberg — Stonyfield Farms
p. 186, Nicola Bullard — Focus on the Global South
p. 189, Farmers Market — Shutterstock image
p. 193, Simran Sethi w/George Siemon — Ethical Markets Media, LLC

Chapter 12

p. 203, Claudine Schneider — Claudine Schneider
p. 205, Dr. Jim Gordon — Center for Mind-Body Medicine
p. 206, Jeff Hollender — Seventh Generation Company
p. 207, Larry Brilliant with Friends in Nepal — SEVA Foundation
p. 211, Paul Freundlich, Simran Sethi, and Dr. Barbara Glickstein — Ethical Markets Media, LLC

Chapter 13

p. 216, Launch of Principles for Responsible Investing — United Nations
p. 217, Nicholas Parker — Cleantech Capital Group
p. 219, Robert Shaw — ARETE Corporation
p. 222, Robert A. G. Monks — Robert A.G. Monks
p. 231, Age of Light — © Hazel Henderson 1991

Index

D

E

F

G

H

I

N

O

P

Q

R

S

T

U

V

W

Y

Z

About the Author

Hazel Henderson, renowned futurist, evolutionary economist, author, and syndicated columnist, created and produces the public television series *Ethical Markets*. A fellow of the World Business Academy, she shared the Global Citizen Award with Nobel Laureate A. Perez Esquivel of Argentina. Henderson's internationally syndicated editorials appear in 27 languages and more than 400 newspapers.

Her articles have appeared in more than 250 journals, including: *Harvard Business Review, Financial Analysts Journal, The New York Times, The Christian Science Monitor,* and *Challenge*. Her books are translated into French, Italian, German, Spanish, Japanese, Dutch, Swedish, Korean, Portuguese, and Chinese. She sits on several editorial boards, including *Futures Research Quarterly, The State of the Future Report, E/The Environmental Magazine* (USA), and *Resurgence, Foresight and Futures* (UK). Henderson is a Patron of the New Economics Foundation (London), an Honorary Member of the Club of Rome, and a board member of Instituto Ethos, Brazil. Her Country Futures Indicators (CFI®), an alternative to the Gross National Product (GNP), is a co-venture with Calvert Group, Inc.: the Calvert-Henderson Quality of Life Indicators (Desk Reference Manual, 2000), updated regularly at www.calvert-henderson.com.

Henderson holds many honorary degrees, is a former advisor to the U.S. Office of Technology Assessment and the National Science Foundation; an active member of the National Press Club (Washington, DC) and the World Future Society (USA); a Fellow of the World Futures Studies Federation; and a member of the Association for Evolutionary Economics. She lives in Florida.